131 WAYS TO USE FILM CONTAINERS

131 WAYS TO USE FILM CONTAINERS

To Teach Literacy, Math, and Science— and Just to Have Fun!

By Donna Whyte

Crystal Springs
BOOKS
A division of SDE Staff Development for Educators
Peterborough, New Hampshire

© 2004 Crystal Springs Books

Printed in the United States of America
 08 07 06 05 04   1 2 3 4 5

Published by Crystal Springs Books
75 Jaffrey Road
P.O. Box 500
Peterborough, NH 03458
1-800-321-0401
www.crystalsprings.com
www.sde.com
ISBN 1-884548-56-3

Art Director, Designer, Production Coordinator: Soosen Dunholter
Photos/Illustrations: Soosen Dunholter
Editor: Sandra Taylor
Project Coordinator: Deborah Fredericks

Dedication

To my sunshine, Taylor, whose creativity and eagerness to be a part of Mom's adventures made many of these ideas come to life.

Acknowledgments

I want to thank Margaret Uebelacker, my friend, my mentor, and a wonderfully gifted teacher. Margaret spent more than 30 years in the classroom, and her enthusiasm for teaching made each year seem like her first. Many of the ideas included in this book were discussed in our "travels" together. Her willingness to listen, create, and add ideas was both inspiring and heartwarming.

Author's Note

As a teacher, I often referred to the children in my classes as Mrs. Whyte's Smarties. Over the years, we created Smarties Publishing for books we authored and The Smartie Press for communicating with families about our classroom activities, and the morning message always began with Dear Smarties. I now have a Web site with this name: TheSmartieZone.com. So, in honor of my smarties, some of the activities in this book make reference to our classroom name.

Contents

Introduction

One year in a kindergarten classroom, I was planning a unit titled "The Sty's the Limit." It was to be based entirely on pigs—the readings, the writing, math, science, and so on—and I knew the kids would love it. I had played a game for adults called Pass the Pigs, which was really fun and kept everyone laughing, but it was too challenging for my young students. So I decided to make up a math game similar to it by using little pigs from my math manipulative kit. Like many of you, I save everything, so I knew that I would find something to put those little piggies in. I went to the closet and located a bag of storage containers. My eyes were drawn immediately to the film containers I had collected. They were the perfect size. I then cut out pieces of brown paper, which I laminated, for "mud puddles" to roll the pigs on, and presto! Piggy in the Mud was born. The scoring system was based on the way the piggy, or piggies, landed (back, side, all fours, or snout—the children's favorite). The game was created with the idea of building number recognition, number sense, and addition. The goal was to have the children count up and record piggy points. We had many great times rolling our pigs and recording our points during center time, math time, and recess.

The next year, some of the children and I continued together to first grade. About a month into the school year, a little girl asked what had happened to our pigs. I told her I had left them in storage in our kindergarten class and would try to locate them. When I found them there in a shoe box, I could not help remembering how much fun they had been. I decided I could tweak the game for first grade by changing the point system and teaching the kids how to skip count when adding their piggy points. The class would once again be learning with a game that kept them laughing.

The following year, we were off to second grade, and Piggy Probability evolved: We graphed the way the pigs landed, based on a specific number of rolls. The game had been not only valuable as a teaching tool but also great fun for the kids. Knowing how small items tend to disappear when working with young children, I always put the game away when we were through playing. In fact, I was quite firm about keeping it in the classroom so that the piggies would not get lost. I guess I must have been pretty tough about that over the years.

At the end of the school year, I decided to give the class a collection of small items that I hoped would remind them of me and the years we had spent together. I called this our class treasure. One of the mementos was a film container with two little piggies inside. I knew the kids would recall the different versions of the game they had played from kindergarten through second grade. As they began removing items from their bags, I heard many recollections of things we had done. When one of the children pulled out the film container, I asked, "Why would this remind you of having had Mrs. Whyte?" The children immediately started naming the games and activities that had involved the piggies.

But one little boy seemed confused. He held up his container and declared, "Someone put this in my bag!" He was afraid that I would think he was trying to take it. I told him, "Sweetie, Mrs. Whyte put it in your bag." He asked if the container was his to keep, and I told him yes, that in this instance, the containers were a gift and were to be kept. He then asked, "But can we take the containers home?" I replied, "Yes, these are for you." He was thrilled. I walked over, dumped the little piggies into my hand, and held them out to him. "You guys just love these little piggies, don't you?" I asked. He looked up and replied matter-of-factly, "No, but I *really* love this little container!"

My first thought was that I had hours of time and money invested in finding and buying the pigs but the containers had been free. Then I realized that if a film container could be *that* exciting to a child, I needed to figure out how many other things I could fit inside one. Well, 131 ideas later, I have an entire book of them to share with you. I hope you enjoy it and have many years of learning fun with your students.

Before You Begin the Activities

The first thing you need to do is collect as many empty film containers as you can. Check with camera shops and any other places that develop film, such as Wal-Mart, Target, and Walgreens. Ask your students and their parents to help with the search.

There are three basic types of film containers—black, clear, and frosted. Most are round in shape, although I have had oval ones, too. The lids are generally black, gray, or clear and vary in how they attach to the container: Some fit inside, flush with the top rim; others snap on over the rim. They also have either a smooth top, a bump in the middle, or a groove around the top.

Each activity has a list of materials, but I have not included things that most of you already have on hand in your classrooms, such as white glue, adhesive putty, a glue gun, an X-Acto or utility knife, tape, and scissors. If an activity requires a certain type of film container, I point this out, but most of the activities can be done with any kind of container or lid. The directions are pretty basic and can be followed word for word or used simply as a guide. If you come up with a better idea for putting something together, by all means do it your way.

Most of the activities can be tiered or leveled, depending on the age and abilities of your students. In some instances, I have suggested ways to simplify them; in others, I have included ways to extend them so they become more challenging. Some activities can be done in groups, some in pairs, and some individually. Some make great center activities, too. In fact, a number of the activities in the Art and Gifts chapters are perfect for centers. It is fun, too, to let the children decide which ones to do. For younger children, you (or a parent or volunteer) will want to do the cutting or other steps that could involve safety issues. Older students should be able to handle most of the activities on their own.

I have included a lot of reproducibles at the back of the book. These will save you time and provide you with handy assessment and self-check tools. Whether you use them is up to you and your classroom needs. In addition to reproducible labels, there are five pages of adhesive labels, at the back also, ready to be peeled off and attached to the film containers. These are another great time-saver.

Here are a few tips to keep in mind as you get started:

1. Before you use the containers, peel off and throw away any labels. Put the containers in a mesh bag and wash them in the dishwasher.*
2. Before you start using the printed adhesive labels, it is a good idea to make photocopies of the pages. That way, you will have extras on hand for later use. (I recommend using #5260 Avery labels or something comparable.)
3. After you attach a label to a container, you might want to cover it with clear tape or Con-Tact paper to protect it from getting dirty or torn.
4. When you have individual containers for your students but the containers do not need labels, put a sticky paper dot on the bottom of each container with the owner's initials written on it. Then you can easily identify which container belongs to which child.
5. When a slip of paper is to be enclosed in a container, it helps to wrap the paper around a pencil to curl it up tightly and then insert it in the container.

*Note: Kodak has received countless queries about the safety of using film containers for purposes other than storing film. The company has stated that there are no toxic residues in Kodak film containers; however, because Kodak does not manufacture these containers in compliance with Food and Drug Administration (FDA) or Consumer Product Safety Commission (CPSC) requirements, it cannot recommend using the containers for anything other than their intended use. Kodak also suggests that, as a safety precaution, the containers and their lids be kept away from small children and pets.

As you do the activities in this book, I'm sure you'll come up with other ways to use film containers. Feel free to share them with me and the editors at Crystal Springs Books by e-mailing us at ideas@crystalsprings.com. If we receive enough suggestions, there could be a volume 2 of this book!

Keep saving and using those containers!

1. Classroom Management

1 Warm Fuzzies

Materials

- Film container of any kind
- Reproducible label: _____'s Warm Fuzzies
- Pom-poms decorated as desired
- Storage container for pom-poms
- Homework Pass reproducible (see page 102)

Directions

1. Keep a large collection of Warm Fuzzies (pom-poms decorated as desired) in a shoe box, metal canister, plastic box, or some other container near your desk.

2. When a child deserves special recognition for a class accomplishment or simply needs a pat on the back for encouragement, give him a film container with a pom-pom inside.

3. After a child receives a certain number of Warm Fuzzies, he can trade them in for a homework pass.

Consider This:

When the class does something as a group that you want to reward them for, give each child a fuzzy.

2 Thinking Lotion

Materials

- Film container of any kind
- Adhesive label: **Thinking Lotion**
- Nice-smelling hand lotion

Directions

1. Fill the container with lotion and keep it at your desk. When a child seems to be struggling for an idea, rub a dab of Thinking Lotion on the child's temple or the tip of her nose.

2. Give her 30 seconds or so to think, and if she still says she cannot think of anything, say, "Oh, it doesn't work for you?" Inevitably, the child will think of something.

Consider This:

Place some beads or small bells inside a film container, add a jewel to the lid, label it "Magic Crystals," and shake it over the child's head. Tell her that the Magic Crystals will help her come up with an idea. Positive thinking can help children overcome mental roadblocks.

3 Encouraging Words

Materials

- Film container of any kind
- Adhesive label: _____'s Cheering for You
- Encouraging Words reproducible (see page 103)

Directions

1. Fill in your name on the adhesive label and attach it to the film container.

2. Cut apart the Encouraging Words and put them all in the container. (Write your own words in the blank grids on the reproducible, and use them instead or in addition to those provided.)

3. When a child is having a difficult time or is feeling discouraged, let him pull out one of the slips of paper and read it for encouragement (or you read it to him, if necessary).

Consider This:

To make the slips longer lasting, laminate the reproducible after cutting it up.

4 Pennies for Patience

Materials

- Round film container for each child
- Reproducible label: _____'s Pennies for Patience

Directions

1. When a child is able to wait patiently, reward her with a penny for her Pennies for Patience container.

2. After she has collected 10 pennies, she can trade them in for a dime and then for a pencil, a bookmark, or something similar.

Consider This:

Every time a child has earned 10 pennies, she can turn them in for a dime and start saving for an ice cream. Later, she can trade 5 dimes for 2 quarters.

5 To Ask or Not to Ask

Materials

- Film container of any kind for each child
- Reproducible label: _____'s Questions: **Choose Wisely!**
- Bingo chips for each container (quantity based on individual child)

Directions

1. Use this activity to discourage the children from asking a lot of unnecessary questions (those that could be figured out on their own or answered by another child in the room).

2. First, decide how many questions you are willing to answer each day (based on what is appropriate for the individual child) and put that number of bingo chips in the child's container.

3. Give the children their containers and tell them that the chips represent the number of questions they can ask you. Be sure to explain that before they come to you with a question, they should first try to figure out the answer themselves or ask another child to answer it.

4. Later, when a child comes to ask you something, say to her, "Is this something you need me to answer?" If she says yes, then listen to the question. If it *was* something you needed to be asked, give the child your answer and allow her to keep all her bingo chips. If, however, the question was one that she could have figured out herself or could have asked another child, tell her that, and ask her to give you one of her chips. Do not answer the question.

5. Remind the children that when they run out of chips, they also run out of questions, so they must choose their questions wisely.

Consider This:

There will always be some children who need to ask more questions than others, so allow more chips for them, then gradually reduce the number of chips they receive.

6 Building Community in the Classroom

Materials

- 5 round film containers
- 5 adhesive labels: Honest, Loyal, Dependable, Responsible, Considerate
- Ruler or paint stick

Directions

1. Glue the 5 containers onto the ruler or paint stick and make sure all the labels face in the same direction.
2. Discuss with the children what "building community in the classroom" means, and brainstorm with them the definitions of the words *Honest, Loyal, Dependable, Responsible,* and *Considerate.*

3. Once you and the children have come up with words that define these, write the words on a sheet of paper, cut them apart, and put the words in the appropriate containers.
4. When the class needs to be reminded of what it means to be *loyal,* for example, open that container and read the words stored inside.

Consider This:

a. To make the individual pieces longer lasting, laminate the sheet of words after cutting it apart.
b. This activity can be reworked so that it applies to other categories, such as Literacy, Math, or Science, and used in a similar fashion. For example, you could change the title to The Water Cycle, use it with words like *Precipitation, Evaporation,* and so on, and make it a Science activity.

7 Good-bye, Homework

Materials

- Film container of any kind for each child
- Reproducible label: _____'s Night Off
- Assortment of gems
- Homework Pass reproducible (see page 102)

Directions

1. Decide ahead of time how many gems will earn a child a night off from homework. Each time a child does his homework, give him a gem to place in his film container.

2. When he earns the required number of gems, reward him with a homework pass.

Consider This:

Base this on homework done on Monday through Thursday. When the child earns gems for each of those days, he gets Friday off.

8 Center Chips

Materials

- Film container of any kind for each child and for each center
- Reproducible labels: _____'s CENTERS: Look What I Finished!
- Different-color chips

Directions

1. Assign a particular color to each center and put chips of the same color at the center in a container. For example, the writing center might be blue, the science center red, and the reading center yellow.

2. Hot-glue a colored chip to the lid of each center's film container.

3. When a child completes a center, she takes a chip from that center and puts it in her film container. At the end of the day, she shows you which centers she completed.

9 Back-to-School Kit

Materials

- Film container of any kind for each child
- Reproducible label: _____'s Back-to-School Kit
- Rules for the Classroom reproducible (right)
- Buttons, Band-Aids, pieces of crayons, rubber bands, erasers, paper clips, Smarties, pieces of aluminum foil, gold ribbons

Directions

1. Do this activity at the beginning of the school year as a way to introduce rules and give visual clues to those rules.
2. Give the children their own film containers and have them keep one of each item inside.
3. Allow the children to take a copy of the rules and their containers home to show their families.

Consider This:

a. Make an enlarged photocopy of the Rules for the Classroom, glue on the real items, and let the children decorate the sheet.
b. Have the children use their crayon pieces to create a rainbow that you can hang by your desk. This will remind the children that they all add a little color to your day.
c. Instead of the piece of aluminum foil, cut out a small mirror shape from oak tag, cover it with foil, and have the children place that in their containers.

d. If your school doesn't allow candy, replace it with stickers or create small "smartie" stickers.

Rules for the Classroom

 Gold ribbon: Friendships are the ties that connect us.

 Paper clip: It's nice to have others to help you hold things together.

 Piece of crayon: Everyone adds a little color to every day.

Button: Button your lip if you do not have something positive to say.

 Piece of aluminum foil (reflective): Treat others as you want to be treated.

Eraser: We all make mistakes.

 **Band-Aid:** To heal hurt feelings.

 Rubber band: Remember to be flexible.

Smartie: Each one of you is my smartie.

10 Nuts & Bolts

Materials

- Film container of any kind
- Adhesive label: NUTS & BOLTS
- 6 sets of nuts and bolts of different sizes, unassembled

Directions

1. Put all the nuts and bolts in the film container and keep it at your desk.

2. If a child needs you while you are busy, hand her the container and tell her to assemble all the nuts and bolts, and that you will be with her as soon as she finishes.

3. Make sure you are able to help her when she finishes. Don't forget her.

Consider This:

This is good for fine motor skills.

11 Home of the "Fit"

Materials

- Film container of any kind
- Adhesive label: Home of the "Fit"
- Small glitter pom-pom with eyes glued on

Directions

1. Use the container to store the pom-pom, called "the Fit."

2. When you want to teach that throwing a fit does not accomplish anything, use the Fit lightheartedly, as you might use a puppet. Do not introduce the Fit when there are problems.

3. In the course of the presentation, throw the Fit and ask your class, "Does throwing a fit solve problems?"

4. Discuss with the children other choices besides throwing a fit.

12 Wanda Whiner

Materials

- 2 clear film containers (1 lid only)
- Reproducible label: Wanda Whiner art
- Colored sand

Directions

1. Use a hole punch to make a hole in the lid. Fill 1 container three-quarters full of sand. Attach the lid to the sand-filled container.

2. Using hot glue, attach the top rim of the empty container to the lid of the other container, creating a timer. Then attach the Wanda Whiner label.

3. Before you have to use this timer, model to the children that a whiner is someone who complains about something but does not offer solutions to the problem, and that the alternative to whining is thinking out a plan.

4. When a child whines to you about something, take out the Wanda Whiner timer. Tell the child to go back to his desk, turn over the timer, and think of some solutions to his problem. When the sand runs out, ask the child to come back to you with his solutions. (He should either stop whining or leave the group.)

Consider This:

a. If you have a child named Wanda in your classroom, be sure to use another name.

b. If you have children who get up and wander around the room, you could make a "Willy Wanderer" timer (see reproducible label, page 159) and use it in the same manner.

13 Tooth Suitcase

Materials

- Film containers of any kind
- Reproducible label: _____'s Tooth Suitcase

Directions

1. When a child loses a tooth, put it in her Tooth Suitcase.

2. At the end of the day, let the child take it home.

Consider This:

Let a child make this as a gift for another child with a loose tooth. Have the child decorate a label, attach it to the container, and tell the child who is receiving it to put his tooth inside, place it under his pillow that night, and wait for the Tooth Fairy to come and take it—and hopefully leave something behind!

14 Change Holder

Materials

- Round film container
- Reproducible label: **Let Me Hold Your Change**

Directions

1. When a child has change from his lunch money, a school program, and the like, store it for him in his own container.

2. At the end of the day, let him take the change container home.

Consider This:

For safekeeping, make a necklace (see Necklace, page 85) for the container and have the child wear it home.

15 Stamp Handle

Materials

- Film container of any kind (lid not needed)
- Small stamp

Directions

For children who have difficulty holding a small stamp without getting covered with ink, hot-glue the stamp to the bottom of the container, which serves as a handle, or use stamps that come with self-sticking backs.

Consider This:

a. Use this same technique with paint sponges.
b. See also Craft Foam Stamps on page 87.

16 First Aid

Materials

- Film container of any kind
- Adhesive label: **First Aid** (or use a character Band-Aid as the label)
- Individual packets of alcohol pads
- Band-Aids
- Q-Tips cut in half

Directions

1. Fill the container with the packets, Band-Aids, and Q-Tips and attach the lid.

2. Make a necklace (see Necklace, page 85) so you can wear it as a "pendant" and take it with you to the playground during recess, on field trips, and on other outings.

17 Project Pieces

Materials

- Film container of any kind
- Reproducible label: **Project Pieces** for _____
- Supplies such as thumbtacks, pushpins, brass fasteners, and paper clips

Directions

Fill the container with supplies to help a child keep track of what he needs when working on a project. For example, if the child is making a mobile, you could precut lengths of string and put them inside the container along with brass fasteners and other supplies for the project.

18 Glue and Spreader

Materials

- 2 film containers of any kind
- 2 adhesive labels: **Glue, Spreader**
- Velcro (self-sticking)
- Q-Tips, cut in half

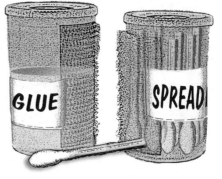

Directions

1. Attach the lids to the containers and velcro the 2 containers together, side-by-side.

2. Fill one container with glue and the other with Q-Tips, and attach the appropriate labels to the containers.

3. Store for future use.

Consider This:

Pull the containers apart to make refilling the glue easier.

19 Ice Pack

Materials

- Film container of any kind
- Adhesive label: **Cool Tube**

Directions

1. Fill the container three-quarters full of water and freeze until needed.

2. When a child has a boo-boo, he can use this instead of going to the nurse's office.

20 Counting Container

Materials

- Film container of any kind for each child
- Reproducible label: _____'s **Counting Container**

Directions

1. Use this container to hold manipulatives for adding and subtracting, or for playing bingo and other similar games.

2. Let the children keep their containers at their desks so you save time by not having to hand them out each time they are to be used.

Consider This:

Use each container to store 10 ones cubes (from math manipulatives) for easy distribution.

21 Emergency Sewing Kit

Materials

- Film container of any kind
- Adhesive label: **Emergency Sewing Kit**
- Buttons
- Needle
- Piece of cardboard that will fit inside container (approximately 1" x 1½")
- 3 pieces of colored thread, each approximately 12" long
- Safety pins
- Needle threader

Directions

1. Hot-glue a button to the lid of the container.
2. Weave the needle through the cardboard and then wrap each length of thread around the cardboard.
3. Place the pins, remaining buttons, thread and needle, and needle threader inside the film container and attach the lid.
4. Keep the kit at your desk for future use.

Consider This:

a. This makes a nice present.
b. If really young children will be making this, be sure to have an adult help them put the needle in place.

2. Literacy

22 Category Sorting

Materials

- 4 film containers of any kind
- 4 adhesive labels: **Storage, Sports, Fruits, Animals**
- Paint stick or large craft stick
- Words and Pictures reproducible (see page 104)

Directions

1. Glue all 4 containers onto the paint or craft stick, spacing them evenly along its length and making sure all the labels face in the same direction.

2. Cut out all the words and pictures on the reproducible. Put them in the Storage container and attach the lid until ready to use.

3. Have a child empty the Storage container and put the words and pictures in the correct category containers: Sports, Fruits, and Animals.

Consider This:

Use a different book title for each of the 3 categories. Write the setting, plot, and characters for each book on a sheet of paper. Cut them out and put them in the Storage container, then have a child sort by book title.

23 Sight Words

Materials

- 24 film containers of any kind
- Reproducible label: **Sight Words Week #_____**
- Sight Words List reproducible (see page 105-107)

Directions

1. Cut out the individual sight words.

2. Put 5 words in each container and attach a label that reads Sight Words Week #_____.

3. Ask a child to take a container to his desk, empty the contents, and alphabetize the words.

Consider This:

It can be difficult for some children to look at words on a wall and then copy them on paper. This activity allows them to have the words right in front of them and on the same plane where they write.

24 Sorting by Colors and Shapes

Materials

- 3 film containers of any type
- 3 adhesive labels: **Words Storage, Color Words, Shape Words**
- Paint stick or large craft stick
- Color and Shape Words reproducible (see page 108)

Directions

1. Glue all 3 containers onto the paint or craft stick, spacing them evenly along its length and making sure all the labels face in the same direction.

2. Cut out the individual words on the reproducible. Put them in the Words Storage container and attach the lid until ready to use.

3. Have a child empty the Storage container and put the words in the correct category containers: Color Words and Shape Words.

Consider This:

Once the words have been categorized, have the child empty the containers and alphabetize the words in both categories. As a self-check, have the child use a copy of the Color and Shape Words reproducible, which is in alphabetical order.

25 Word Families

Materials

- Film container for each word family
- Reproducible label: **Word Family**
- Word Families reproducible (see page 109)
- Alphabet I reproducible (see page 109)
- Real Words/Non-sense Words reproducible (see page 110)

Directions

1. Choose one 2-letter word family from the Word Families reproducible and any number of consonants from the Alphabet I reproducible and enclose them in that word family container.

2. Have a child empty the container, place a consonant next to the word family, and write that word under the appropriate heading on the Real Words/Nonsense Words reproducible.

3. Extend the activity by adding other word families that are not included here.

Consider This:

Be sure to go over the list of words with the child to see what he understands to be real or nonsense words.

26 Guess My Letter

Materials

- Black film container
- Adhesive label: **Guess My Letter**
- Alphabet II reproducible (see page 111)
- Guess My Letter reproducible (see page 112)

Directions

1. Cut out the letters of the alphabet and put them in the container.

2. Pull out a letter and give a child clues about it. If, for example, the letter is *m,* you might say, "I am a letter near the middle of the alphabet. I am found in the word *monkey.* I make this sound: [make the *m* sound]." Write in your own clue for the 4th guess.

3. If necessary, continue to give clues until the child guesses the letter.

1st Guess ☐ I am between __K__ and __P__ in the alphabet.
2nd Guess ☐ I am found in the word __Monkey__ .
3rd Guess ☐ I make this sound: _____
4th Guess ☐ _____

Consider This:

Let 2 children do this together: One gives clues and the other guesses; then they swap roles.

27 Finger Puppet Characters

Materials

- Film container of any kind for each child
- Cotton balls
- Craft foam
- Construction paper

Directions

1. Provide each child with a film container plus pieces of craft foam and construction paper to cut up and glue to the container.

2. Have the children make finger puppet characters that match those in a book you are reading to the class.

3. Insert a cotton ball into the container so it will not slip off small fingers.

4. Have the children act out the story with their puppet characters.

Consider This:

Allow the children to make up their own characters and write stories about them later.

28 Story-in-Waiting

Materials

- 3 film containers of any kind
- 3 adhesive labels: **Setting (Where?), Action (What?), Characters (Who?)**
- Paint stick or large craft stick
- Setting, Action, and Characters reproducibles (see pages 113–115)
- Different-color paper for photocopying

Directions

1. Glue all 3 containers onto the paint or craft stick, spacing them evenly along its length and making sure all labels face in the same direction.

2. Photocopy the reproducibles on different-color paper to make sorting easier. Cut out the words and then put them in the appropriate containers.

3. Allow each child to select a word from each container and then do one of the following:

    a. Write a sentence based on the words picked

    b. Write a story based on the words picked

    c. Draw a picture based on the words picked

Consider This:

a. Choose a word from each category, put the words in a film container, and label it "Story-in-Waiting." Repeat with other containers, then put all the containers in a box or basket. Have each child pick a container and write a sentence or story or draw a picture based on the words inside.

b. Instead of words, place a collection of small items (such as a button, a penny, and a ring) in a container and ask a child to write a sentence or story that connects the items.

29 Feel Me!

Materials

- Film containers of any kind
- Variety of small items, such as aluminum foil, bubble wrap, cotton balls, or a pebble— anything that matches the descriptive words you are teaching

Directions

1. Place each item in a film container and number the container.

2. Ask a child to open a container and write 3 to 5 descriptive words for the item inside. For example, if the container has a cotton ball, the child might write "soft," "light," and "white."

Consider This:

To get the children to rely on their sense of touch, have them do this activity with a partner and with their eyes closed or their hands behind their backs so they cannot see the item placed in their hands. Using half of the containers, one child gives his partner an item and writes down the descriptive words the partner says. Then they switch roles, using the remaining containers.

30 Contractions

Materials

- 7 film containers of any kind
- 7 adhesive labels: Words with "am," Words with "are," Words with "is," Words with "have," Words with "will," Words with "would," Words with "not"
- Contractions reproducible (see pages 116–118)

Directions

1. Photocopy the reproducible on different-color paper (the words on one color, the contractions on another) and cut apart.

2. Put the words and contractions in the appropriately labeled film containers.

3. Have a child empty a container, match the words to the contractions, and then record these on a sheet of paper.

Consider This:

a. Eliminate the color hint for second graders.
b. As a self-check, give the child the Contractions reproducible to refer to.

31 Punctuation Marks

Materials

- Film container of any kind
- Adhesive label: **Punctuation Marks**
- Punctuation reproducible (see page 119)
- Adhesive putty

Directions

1. Cut out individual punctuation marks and put them in the film container.
2. Put a piece of adhesive putty on the inside of the container lid.
3. During Morning Message or when the class is working on adding punctuation to interactive charts, poems, and the like, have a child pinch off a tiny piece of the adhesive, attach it to the back of the correct punctuation mark, and place the mark where it belongs.

Consider This:

Begin with periods and question marks for kindergartners, then add commas, quotation marks, and exclamation marks.

32 Pet Rock Home

Materials

- Film container of any kind for each child
- Reproducible label: _____'s **Pet Rock**
- Rocks small enough to fit in container
- Paint or colored markers

Directions

1. Set out the rocks and allow each child to select one. (Or let each child bring in a small rock from home or recess. The children should take their containers with them to be sure their rocks fit inside.)
2. Ask the children to decorate their pet rocks as they desire, using paint or markers.
3. Ask each child to draw a picture or write a story featuring his pet rock.
4. When the children have finished, remind them to return their pet rocks to their containers and to save them for future adventures.

Consider This:

Let each child name his pet and add that to the label.

33 Rebus Story

Materials

- Film containers of any kind for each child
- Reproducible label for each container: **Rebus Story**
- Rebus Symbols reproducible (see page 120)

Directions

1. Cut out the rebus symbols and place 3 to 5 in each film container, depending on the age of the children.

2. Have each child choose a container and, after looking at the contents, write a story and glue the rebus images within the text.

went to the farm and saw a and a and a . Then they picked s. It was a day. They had some lemonade and .

Consider This:

To increase the difficulty, increase the number of rebuses in each container.

34 Reading Robot

Materials

- Film container of any kind for each child
- Reproducible label: _____'s **Reading Robot**
- Reading Robot reproducible (see page 121)

Directions

1. Give each child a film container with her name written on the label.

2. After the child reads a book, have her fill out a Reading Robot form, roll it up, and put it in her container.

Title _____
Date Completed _____
Book Number # _____

Consider This:

a. Send the child's robot home with her so that her parents can see what books she has read.

b. Give children who are too young to fill out the form a bingo chip for each book to keep in their robots.

c. Make a large robot out of a coffee can for storing the slips and chips.

35 Rhyming Jars

Materials

- 4 film containers of any kind
- 4 reproducible labels: **Rhymes with** _____ (3), **Words**
- Paint stick or large craft stick
- Blank Grid reproducible (see page 122)

Directions

1. Choose 3 words to write on the Rhymes with _____ labels (for example: cat, dog, tin).

2. Glue all 4 containers to the paint or craft stick, spacing them evenly along its length and making sure all the labels face in the same direction.

3. On the reproducible, write words that rhyme with the label words (for example: sat, rat, mat; fog, hog, log; bin, win, pin). Photocopy the page, cut the words out, and place them in the Words container.

4. Have a child empty the Words container and put the words in the correct rhyming word containers.

Consider This:

a. After you fill out the reproducible grid, make an extra copy of it for the children to use as a self-check.

b. For younger children, this activity could be done with pictures instead of words.

c. Use with nursery rhymes that have strong word families, such as "Jack and Jill."

36 Building Words

Materials

- 3 film containers of any kind
- 3 adhesive labels: **Consonants** (2), **Vowels**
- Paint stick or large craft stick
- Alphabet I reproducible (see page 109)
- Real Words/Nonsense Words reproducible (see page 110)

Directions

1. Glue all the film containers to the paint or craft stick, spacing them evenly along its length and making sure all the labels face in the same direction.

2. Using the Alphabet I reproducible, photocopy the vowels and consonants on different-color paper and cut them out.

3. Place the vowels in the Vowels container and the consonants in the Consonants containers. (You do not have to use all the vowels and consonants.)

4. Have a child remove 1 letter from each container and arrange the letters to make a 3-letter word.

5. Using the Real Words/Nonsense Words reproducible, have the child write the word under the heading he feels is correct.

. . . .

Consider This:

a. This is a great letter-blending activity.

b. For younger children, you may want to provide only 1 vowel for them to keep reusing.

37 Brain Freeze

Materials

- Black film container
- Adhesive label: **Brain Freeze**
- Words printed on slips of paper or 3 to 7 small items that will fit in container

Directions

1. Place words or items in the container.

2. Toss the contents onto a surface and let the children look at the collection for 10 to 20 seconds.

3. Put the contents back in the container and ask the children to name or write as many of the words or items as they can remember.

Consider This:

a. Good for oral language development and for learning to categorize for memory skills.

b. Introduce this using only 3 items and gradually increase the number.

c. If using words, write the same word on both sides of the paper so you do not lose time turning over the slips.

d. After playing as described above two or three times in a row, and using at least 7 items, make the activity more challenging by removing 1 item or word before tossing out the contents and see if the children can discover what is missing.

38 How Are We Alike?

Materials

- Black film container
- Adhesive label: **How are we alike?**
- 4 or 5 small pictures or items that have some characteristic in common (items of the same shape or color; items that are eaten; or items that hold things together, such as a staple, tape, brass fastener, chicken ring)

Directions

1. Put the pictures or words in the container.

2. Have a child remove 2 items and write a list of ways they are alike. For a younger child, you can write down what the child says.

3. Have the child remove another item, place it with the others, and see how the 3 items are alike. The child may need to cross out some characteristics written earlier and add others to the list.

4. Repeat until the child has removed all the items from the container and edited the list to include only those ways in which all the items are alike.

Consider This:

This works well as a paired activity, where one child removes the items from the container and another writes down the ways the items are alike.

39 Word Play

Materials

- Film container of any kind
- Adhesive label: **Word Play**
- Word Play reproducible (see page 123)

Directions

1. Write a word with at least 4 letters on a sheet of paper and cut it apart into individual letters.

2. Put all the letters in the film container.

3. Give a child a copy of the Word Play reproducible and the film container.

4. Have him empty the container and use the letters to make as many 2-letter, 3-letter, and 4-letter words as possible, recording each one on the sheet.

Consider This:

a. Visit TheSmartieZone.com, click on the fish, scroll down to Word Play, and then click on Words in a Word for easy access to words and games.

b. Cut vowels out of a different color for younger children.

c. Use alphabet macaroni instead of cut-up letters.

d. Use the overhead so the entire class can work at the same time on the same word.

e. Give each child a film container with letters to work on at home.

40 Open House Scavenger Hunt

Materials

- Film container of any kind for each child
- Reproducible label: _____'s **Scavenger Hunt**
- Scavenger Hunt reproducible (see page 124)

Directions

1. Prior to Open House, ask the children to decide what they want their family members to search for during their school visit and establish some ground rules. For example, do the items need to be within the classroom? Does the "hunter" need to pick up the item, or can he simply check each off the list when found?

2. Have each child write her name on a copy of the Scavenger Hunt reproducible, list the items to be found, roll the slip up, and place it in her container.

3. At Open House, ask the child's family member(s) to write a short note to the child on the back of the slip, describing how the search went, and place it inside the container. The next school day, give the child her container with the note to read.

Consider This:

a. The children can test how clear their ground rules are by exchanging containers with each other prior to Open House.

b. Ask volunteers to help you complete the scavenger hunt for a child whose family did not attend Open House.

Open House
Scavenger Hunt _____
_____
Find these items in _____
our classroom: _____

41 Getting to Know You

Materials

- Film container of any kind for each child
- Reproducible label: **All About**

Directions

1. Give each child his own container with his name on it and ask him to fill it with items or pictures that describe who he is.

2. As a speaking activity, ask each child to share with the class why he chose these items for his container.

Consider This:

a. Have each child write a story about himself based in part on the items from his container.

b. Have the children trade containers and write stories based on the items they get.

c. Have the children compare and contrast the information in their containers.

42 Message in a Container

Materials

- 3 film containers of any kind
- 3 adhesive labels: **Help! Open Me!**
- Message in a Container reproducible (see page 125)
- Pom-pom with eyes glued on, or some other figure small enough to fit in container

Directions

1. Write your name on each message slip. Enclose a slip and a pom-pom or small figure in each container.

2. Leave the containers in different locations at school, such as the nurse's office, library, cafeteria, or teacher's mailbox, and wait for them to be returned to your classroom.

3. Share the returned messages with the class either by reading them yourself or by having a child read them.

Consider This:

Write down when and where you placed the containers so the class can keep track of how long it took for them to be found. You also could have the children make predictions and compare them to the actual times.

43 Guess What's Inside

Materials

- Black film containers
- Reproducible label: **What am I?**
- Items small enough to fit in container (examples: penny, marble, button, paper clip, eraser)

Directions

1. Put 1 item in each container.
2. Write 3 hints as to the identity of the item on a reproducible label and tape the label to the container.
3. Number each label so that the children can guess the contents of more than one container.
4. Ask the children to write their guesses on a numbered sheet of paper that corresponds to the numbered labels.

Consider This:

a. Using clues to figure out what is inside a container strengthens high-level thinking skills.

b. Send an empty container and a blank label home with each child. Have the child find a small item to put inside, write hints about it on the label, and attach the label to the container. Then all the children can bring their containers to school so that others can guess the items inside.

44 My Mini Mementos

Materials

- Film container of any kind for each child
- Reproducible Label: _____'s Mini Mementos
- Craft supplies

Directions

1. Have the children create a mini version of themselves using a film container and craft supplies.
2. At the end of the school week, have each child write down something he remembers from that week, roll up the slip of paper, and store it in his container.
3. Do this from time to time throughout the year, then let the children read their memories to the class at the end of the year.

Consider This:

a. This does not have to be done every week—just when it seems appropriate or when you might need a short filler.

b. Let the children write about anything school- or family-related.

45 Giraffe, Monkey, and Chicken Sort

Materials

- 4 film containers of any kind
- 4 adhesive labels: a giraffe, a monkey, a chicken, **Letters**
- Paint stick
- Alphabet III reproducible (see page 126)

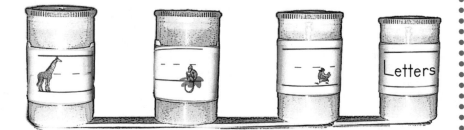

Directions

1. Glue all 4 containers onto the paint stick, spacing them evenly along its length and making sure all the labels face in the same direction.

2. Cut out the letters of the Alphabet III reproducible and put them in the Letters container.

3. Have a child empty the Letters container and put each letter in the appropriate container—giraffe, monkey, or chicken—depending on

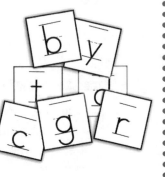

whether the letter touches the top and bottom lines (giraffe), hangs below the line (monkey), or never crosses the road (chicken).

Consider This:

Ask a child to alphabetize the contents of each container and to check herself by using a copy of the Alphabet III reproducible.

46 Ask Me About . . .

Materials

- Film container of any kind for each child
- Reproducible label: **Ask Me About** . . .
- Ask Me About . . . reproducible (see page 127)

Directions

1. Photocopy the Ask Me About . . . reproducible, fill in the "Ask about" line on each strip (referring to something the children did in school that day), and photocopy the sheet again so that you have enough strips to send one home with each child.

2. Have each child write her name on the strip, roll it up, and place it in her container.

3. Tell the children to take their containers home and have a family member ask them about the lesson or event. Have the family member record the child's answer on the form, then send the form back to school in the container the next day.

4. Read each child's form to find out what she got out of the previous day's lesson or event—or perhaps what she *did not* get out of it.

Consider This:

a. Serves as an assessment of a child's retention and ability to communicate details and is great as a home–school connection.

b. To make sure the container is noticed by parents, turn it into a necklace (see Necklace, page 85) and have the child wear it home and then back to school the next day.

Name _____ Ask about _____

_____

_____

Please write your child's response on the lines provided and return it in the container to school.

47 Filmstrip Writing

Materials

- Film container of any kind for each child
- Reproducible label: _____'s Filmstrip
- Filmstrip Writing reproducible (see page 128)

Directions

1. Give each child his container.

2. Photocopy the Filmstrip Writing reproducible and cut the copies into individual strips.

3. Let the children take as many strips as they want, tape them together end to end, and then create a filmstrip using words, drawings, or a combination of both. (Make sure they put tape on the back side of the strip so it doesn't cover up the drawing area.)

4. When the children are finished, have them roll up their filmstrips so that the end of the story is on the inside of the roll. Then have them place the filmstrips in their containers.

Consider This:

a. A fun addition is to cut a vertical slice in the container so that 1 frame can be pulled out at a time.

b. For young children, enlarge the size of the filmstrip and use a small Pringles can instead of a film container.

48 What If . . . ?

Materials

- Film container of any kind for each child
- Reproducible label: **What If . . . ?**
- What If . . . ? reproducible (see page 129)

Directions

1. Photocopy the What If . . . ? reproducible, cut it apart, and put 2 or 3 slips inside each container.

2. Put the containers in a bowl or basket and let each child pick one.

3. Have the children take their containers to their seats, select one of the strips, and write about it.

Consider This:

Be sure to include some silly questions, such as "What if penguins could fly?" or "What if it really rained cats and dogs?"

49 Book Review

Materials

- Film container of any kind for each child
- Reproducible label: **Book Review of**
  _____
- Book Review reproducible (see page 130)

Directions

1. Put a number of film containers and blank labels in a small box or basket in a central location or near where you keep books in your classroom.

2. Make photocopies of the Book Review reproducible, cut them apart, and stack them next to the box or basket with the empty film containers and labels.

Book Review	Reviewer's name:_____
	Title:_____
	Author:_____
	Why I like or dislike this book:_____
	_____

3. After a child has read a book, have her fill out a review.

4. Have her put her review in a film container, write the book's title on a label, attach the label to the container, and put the container in a separate box or basket next to the first one.

5. Each time a child reads a book, she fills out a review. If there is already a container that is marked with the book's title, the child puts her review in that container. If there is no container

for the book she has read, she labels a new container and puts it in the box or basket with the other labeled containers.

Consider This:

a. Eventually, the containers will have reviews from various children, so when a child is trying to decide what book to choose, he can read the reviews to see what others thought of the books he is considering.

b. To keep this organized, find a shelf or surface where labeled containers can be stored and occasionally have a child alphabetize the containers by title so it is easier to look up the book reviews.

c. For fun, you might decorate each film container with a bookworm made out of a pipe cleaner and 2 beady eyes.

50 Sorting by Syllables

Materials

- 4 film containers of any kind
- 4 adhesive labels: Words Storage, 1-Syllable Words, 2-Syllable Words, 3-Syllable Words
- Paint stick or large craft stick
- 1-, 2-, and 3-Syllable Words reproducible (see page 131)

Directions

1. Glue all 4 containers onto the paint or craft stick, spacing them evenly along its length and making sure they are in order and all the labels face in the same direction.

2. Cut the words out of the reproducible and put them in the Words container.

3. Ask a child to empty the Words container and put the words in the containers labeled with the correct number of syllables.

. . . .

Consider This:

a. After the child has sorted the words, ask him to alphabetize them and check his work by referring to a copy of the 1-, 2-, and 3-Syllable Words reproducible, which is in alphabetical order.

b. Use the Blank Grid reproducible (see page 122) and write in your own choice of words for this activity. Be sure to alphabetize them vertically.

3. Math

51 Numeral-Word Match

Materials

- Film container of any type
- Adhesive label: **Numeral-Word Match**
- Numerals and Number Words reproducible (see page 132)

Directions

1. Cut the reproducible into individual numerals and number words and place them in the container.

2. Have a child empty the container, match the numeral to its number word, and record the matches on a sheet of paper.

Consider This:

a. Do a similar activity early in the school year to help classmates learn each other's names. Using the Blank Grid reproducible (see page 122), write each child's first and last name on the grid, cut out all the names, and place them in a film container. Have a child empty the container, match the first and last names, and copy them onto a sheet of paper.

b. Use the Blank Grid reproducible for matching colors to color words or months to their abbreviations.

52 Measure Up!

Materials

- Film container of any kind
- Adhesive label: **Measure Up!**
- Cloth or plastic measuring tape
- Piece of string of any length
- 12 paper clips

Directions

1. Put the rolled-up measuring tape, string, and paper clips linked together inside the film container.

2. Give a child the container and ask her to go around the room and measure different things with the measurers, then record what she finds. For example, she might write or say, "The pencil is ____ paper clips long." "The table is ____ inches high." "The classroom door is ____ strings wide."

53 Piggy in the Mud

Materials

- Film container of any kind
- Adhesive label: **Piggy in the Mud**
- Piggy in the Mud reproducible (see page 133)
- 2 pigs small enough to fit easily in container
- Black or brown paper cut in the shape of a mud puddle

Directions

1. Place the pigs in the container. Decide how many points each pig position will receive and mark this on the Piggy in the Mud reproducible. Photocopy the reproducible and give each player a copy.

2. Give the players a paper mud puddle to roll the pigs onto and the container with the pigs inside, and describe how to play the game.

   a. One child rolls the pigs like a pair of dice.

   b. Depending on how the pigs fall, the child gets so many points and writes that number on his score sheet.

   c. The next child rolls the pigs and marks her points.

   d. Once one of the 4 columns is full on any player's score sheet, the game stops.

   e. Each child adds up his piggy points in each column, then adds up the column totals to get his total piggy points. The player with the most points wins.

3. Remind the children to count up or skip count when adding their totals.

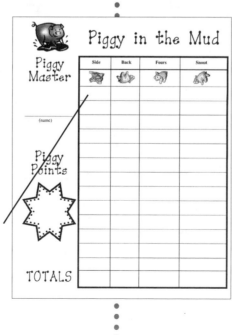

	Side	Back	Fours	Snout
Piggy Master				
(name)				
Piggy Points				
TOTALS				

Consider This:

By changing the point value, you can alter the difficulty of this activity.

54 Guess the Pattern

Materials

- Black film container
- Small stickers or rubber stamps (no wider than 1½")
- Strip of white paper approximately 1½" x 12"

Directions

1. With a knife, make a vertical slit 1½" long in the side of the film container.

2. Attach stickers to the strip of paper or use rubber stamps to create a pattern on it. Leave about ½" between the stickers or stamped images. (A simple pattern might be square, star, star, square, star, star, square, star, star, and so on.)

3. At the end of the strip where the pattern starts, attach a tab that is slightly larger than the container slit so that the strip does not slip back inside. Write "Guess the Pattern" on the tab.

4. Insert the other end of the strip into the vertical slit, feeding the paper in slowly so that it rolls smoothly into the container.

5. Have a child slowly pull the strip out and try to guess what the next shape might be.

Consider This:

a. If you make several containers with different patterns, number the "Guess the Pattern" tabs. Have each child who uses them write the numbers on a sheet of paper and draw how she thinks each pattern would look if she continued unrolling the strip.

b. Give the children blank strips of paper and have them create patterns for the containers.

c. Instead of using various shapes, repeat the same shape and have a child color them to create a color pattern.

d. Have upper-level (second-grade) children create number patterns, such as 5, 10, _____, 20 or 3, 6, _____, 12.

55 My Time Line

Materials

- Round black film container for each child
- Reproducible label: _____'s Time Line
- Time Line reproducibles (see page 134)

Directions

1. Write each child's name on a reproducible label and attach it to a container.

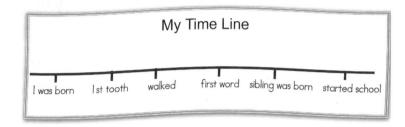

2. Photocopy the My Time Line reproducible, cut it apart, put one in each container, and give each child his container.

2. Send the containers home with the children, along with a note to parents asking them to help their children fill in the dates for the events on the time line.

3. Have the children bring back their containers and use the information gathered to create graphs and charts of their personal data.

My Time Line

My Time Line

I was born   1st tooth   walked   first word   sibling was born   started school

Note: If your child progressed in a different order, mark through the above and write the events in chronological order.

Consider This:

a. Share the graphs and charts with the children and their parents at Open House to illustrate that we all develop at different rates.

b. To find out what the children learned in school each day or week, fill in one of the blank time lines regarding a particular lesson or activity.

56 Sort by Attributes

Materials

- Film container of any kind
- Adhesive label: **Sort**
- Pennies, buttons, beads, or pasta of various shapes, colors, sizes, and so on

Directions

1. Put some pennies, buttons, beads, or pasta inside the film container.

2. Give a child the container and have her sort based on the items enclosed, but do not give her any hints or directions as to how to sort. (Higher-level thinking is required to decide how to sort without directions.) For example, pennies may be sorted by shininess or dullness, or by decades (1980s, 1990s, and so on); buttons, by size, shape, color, or number of holes; beads, by size, shape, or color; and pasta, by size, shape, or color.

Consider This:

a. If you are doing this as a classroom lesson, give a group of children some of the items and tell them to sort the items however they wish, then compare how the items were sorted.

b. As a center activity, fill multiple containers with different items for sorting. After sorting, graph the results to show comparisons.

57 Bead Patterns

Materials

- Film container of any kind
- Adhesive label: **Build a Pattern**
- Length of string, knotted at one end so beads will not fall off
- 10–15 colored beads, at least 3 or 4 of each color

Directions

1. Put the string and loose beads in the container.

2. Ask a child to string the beads in a particular pattern and then draw and color the same pattern on a sheet of paper.

Consider This:

a. Instead of string, let a younger child use a pipe cleaner, which is easier to get the beads on.

b. Ask the child to create a pattern using 3 different colors of beads and to duplicate that pattern on paper using the letters *A, B,* and *C.*

c. Include a slip of paper in the container illustrating some pattern, such as AABCAABC, and have the child reproduce this with the colored beads.

61 Fact Family Game

Materials

- Film container for each fact family
- Reproducible label: **Fact Family** _____ (fill in a number)
- Dried lima beans, pennies, or similar items
- Fact Family Game reproducible (see page 137)

Directions

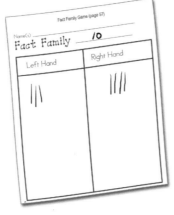

Fact Family Game (page 57)

Name(s) _____
Fact Family _____ 10 _____

Left Hand	Right Hand
|||	||||

1. If using 10 as the "Fact Family," put 10 lima beans or other items in the film container.

2. With the children working in pairs, have one child empty the container, divide the items between his hands, and put his hands behind his back. The other child chooses a hand and tries to guess how many items are in it. As the second child guesses, the one holding the items can help by saying "higher" or "lower." Once the second child makes the correct guess, the first child shows her the contents of that hand and writes that number in the appropriate column of the reproducible.

3. Then the second child tries to figure out how many items are in the first child's other hand. When she gives the correct number, he writes it in the other column of the reproducible.

4. The two then swap places and repeat the game.

62 Estimate

Materials

- 4 black film containers (use clear ones for younger children)
- 4 adhesive labels: **Estimate #1, Estimate #2, Estimate #3, Estimate #4**
- Popcorn kernels, dried lima beans, rice, and dried macaroni (or other items of your choice)

Directions

1. Put any number of one item in each of the 4 containers.

2. Ask the children to estimate how many items are in each container and to record their answers on a sheet of paper.

3. Gather the children's papers and graph their answers to show how they range.

Consider This:

a. Be sure to count the items in each container so you will know ahead of time what the totals are.

b. After all the children have had a chance to record their estimates, use this information for a class discussion on how the size of the items affects the number that fit inside the containers.

c. Create a number line for each item that includes lowest to highest estimates. Mark the children's estimates with blue dots and the actual number of items with red stars.

63 Box Game

Materials

- 2 film containers of any kind
- 2 adhesive labels: MAKE A BOX GAME
- 50 1" pieces of Wikki Stix*
- 30 chips (15 of one color, 15 of another color)
- Make a Box Game reproducible (see page 138)

Directions

1. Put 25 pieces of Wikki Stix and 15 chips of one color in one of the containers. Put the remaining half of Wikki Stix pieces and chips in the other container.

2. With 2 children playing the game, give each a film container. Have them take turns laying 1 piece of Wikki Stix at a time on the dotted lines of the reproducible, connecting the dots. The goal is to make it difficult for your opponent to complete a box, while you try to make as many boxes as possible.

3. When a player completes a box, she places one of her chips in the center of it. When all the boxes have been completed, the winner of the game is the one with the most chips in the boxes.

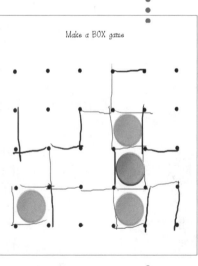

Make a BOX game

Consider This:

a. Laminate the reproducible to make it more durable.
b. Try this game using pipe cleaners instead of Wikki Stix.

Wikki Stix are available through the Crystal Springs Books mail-order catalog. To obtain a copy, call 1-800-321-0401 or go to their Web site: www.crystalsprings.com.

64 Jumping Frog

Materials

- Film container of any kind for each child
- Reproducible label: _____'s **Frog**
- Frog small enough to fit in container for each child
- Craft stick, paint stick, or ruler
- Rubber band

Directions

1. Put a frog in each container and give each child her container.

2. Assemble a frog-jumping device as follows: Set a film container on its side and lay the stick across the container, creating a seesaw effect. Loop the rubber band over the top of one end of the stick. Pull it down and around the film container, then up and over the opposite end of the stick (see illustration).

3. Have a child place her frog on one end of the stick and push down quickly on the opposite end to make the frog jump.

4. Mark where the frog lands. Continue this with the rest of the children, then measure to see which frog went the farthest.

Consider This:

Move the container under the stick from the middle and toward one end of the stick to determine which position enables the frogs to jump the farthest.

65 Weigh Me!

Materials

- 5 film containers of any kind
- 5 adhesive labels: **Rice, Popcorn, Cotton Balls, Pennies, Cheerios** (or Beans, see extra adhesive label provided)
- Rice, popcorn kernels, cotton balls, pennies, and Cheerios (or other items of your choice, such as beans)
- Weigh Me! reproducible (see page 139)

Directions

1. Fill one container with rice, another with popcorn, and so on. Attach the appropriate labels.

2. Tell the children to write what they predict the weight of each container to be on the Weigh Me! reproducible *without* handling the containers.

3. Then have them figure out the weight of each container by using a scale or simply by holding each container in their hands. Tell them to record the results on the reproducible.

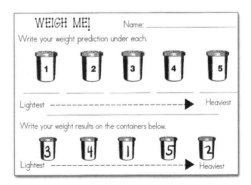

WEIGH ME!   Name: _____
Write your weight prediction under each.

1   2   3   4   5

Lightest ----------------------> Heaviest

Write your weight results on the containers below.

3   4   1   5   2

Lightest ----------------------> Heaviest

66 Coin Count

Materials

- Film container of any kind
- Adhesive label: **Coin Count**
- Assortment of pennies, nickels, dimes, and quarters (real or play money)

Directions

1. Fill container with an assortment of coins.

2. Ask a child to determine the value of the coins in the container.

Consider This:

a. For a greater challenge, have the child use the coins in the container to find 3 different combinations of coins to equal a predetermined amount, such as 57 cents.

b. Ask the child to illustrate her answers by using money stamps or by tracing around the coins and putting a *p, n, d,* or *q* inside each coin to represent penny, nickel, dime, and quarter.

67 Dice Game

Materials

- Film container of any kind
- Adhesive label: **Dice Game**
- Pair of dice
- Hundreds Chart reproducible (see page 136)
- 2 different-color pens or pencils

Directions

1. Place the dice in the film container and give the container to 2 children. Have each roll the dice from the container. The one with the higher total gets to go first.

2. The first player rolls the dice, adds the 2 numbers together, and moves that number of times on the reproducible. She then marks the spot with an X, using a colored pen or pencil.

3. The other child does the same thing but marks his spot with a different-color pen or pencil.

4. The game continues until one child reaches or passes the 100 square.

Consider This:

a. Change the game so that the winner must land exactly on the 100 square. If one player's final roll does not produce the count needed, he does not move forward, and the other player gets a turn. This continues until one player rolls the required count. (If a player lands on 99, he then rolls only 1 of the 2 dice.)

b. To accommodate lower grade levels, use only 1 die.

68 Penny Toss

Materials

- Film container of any kind
- Adhesive label: **Penny Toss**
- Penny
- Penny Toss reproducible (see page 140)

Directions

1. Place the penny in the film container, snap on the lid, and give it to a child along with a copy of the Penny Probability reproducible. (Do not forget to fill in the number of tosses.)

2. Have the child fill out the top portion of the reproducible.

3. Ask the child to shake the container, then shake out the penny and record whether it landed heads or tails up. Have the child repeat this for the specified number of times and tally the results.

4. Use the results for a class discussion on probability. Based on the results, ask the children how many times they think it would land heads up versus tails up. Ask them if they think things would change if another penny was added.

Consider This:

Using 2 pennies, have the children tally the results for the times both coins landed heads up, both landed tails up, and one landed heads up and the other tails up.

69 Fraction Fun

Materials

- Film container of any kind
- Adhesive label: **Fraction Fun**
- 10 dried lima beans or similar items
- Nail polish, nontoxic spray paint, or permanent marker

Directions

1. Color one side of each lima bean using nail polish, nontoxic spray paint, or permanent marker. Let the beans dry, then place them in the film container.

2. Have a child toss the beans from the container and record the number of colored sides as compared to the total number of beans. For example, if 10 beans are tossed and 3 beans land with the colored side showing, the child would write "3 out of 10."

3. Have the child continue to toss the beans a specified number of times and record the results in this manner (1 out of 10, 5 out of 10, and so forth) until they have a thorough understanding of fractions.

Consider This:

Change the number of beans to work with different fraction groups.

70 Cross It Out

Materials

- Film container of any kind
- Adhesive label: **Cross It Out**
- Two dice
- Cross It Out reproducible (see page 141)

Directions

1. Place the dice in the film container and give the container to 2 children. Give both a copy of the reproducible. Have each roll the dice from the container. The one with the higher total gets to go first.

2. The first player rolls the dice and decides how he wants to cross off the numbers rolled. For example, if the dice rolled are 6 and 2, he could cross out the actual numbers (6 and 2), cross out 8 (the sum of the numbers), or cross out 4 (the result of subtracting the numbers). The only way he can cross off 0 is if he throws a double: $4 - 4 = 0$. Remind the children that they can cross off only one of these choices.

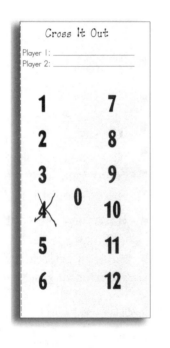

3. Once the child crosses off a number on his sheet, he cannot use it again. Sometimes the rolled dice will not produce any new numbers to cross off.

4. The child who crosses off all the numbers first is the winner.

Consider This:

a. You can also specify a certain number of rolls. Then the children add up the numbers that are not crossed off their sheets, and the one with the smaller number wins the game.

b. Another quick game can be played by allowing the children to cross off all numbers that the dice can represent (that is, the actual numbers, the sum of the numbers, and the result of subtracting the numbers).

4. Science

71 Liftoff!

Materials

- Clear film container with lid that fits inside and flush with top rim of container rather than over top rim
- Reproducible label: rocket wrap art
- Empty toilet paper roll
- Clear Con-Tact paper
- Styrofoam plate
- Alka-Seltzer tablets

Directions

1. Have a child color and decorate the toilet paper roll (the launcher).
2. Attach the label to the film container (the rocket) and cover with Con-Tact paper.
3. Cut four to six 1" vertical slits equally spaced around one end of the toilet paper roll to create "launcher legs." Bend these legs outward and tape them to the styrofoam plate (the launchpad).
4. Pour 1/8 cup (2 tablespoons) water into the rocket, add ¼ Alka-Seltzer tablet, and quickly snap in the lid. (Lid should be secure so that you can't hear air escaping.)
5. Place the rocket, lid end down, into the launcher. Tilt it away from the children and wait . . . for . . . liftoff! (There are 30–45 seconds to liftoff—time for a child to secure the lid.)

6. After the rocket lands, measure the distance traveled and discuss what propelled the rocket.

Consider This:

a. It is wise for the "launch commander" to wear safety glasses during the procedure.

b. The rocket can be colored with permanent marker instead of covered with the label.

c. Increase the "fuel" to ½ tablet and have the children predict whether this will affect the distance the rocket will travel.

d. Don't substitute a paper plate for the styrofoam plate because it won't hold up to the test.

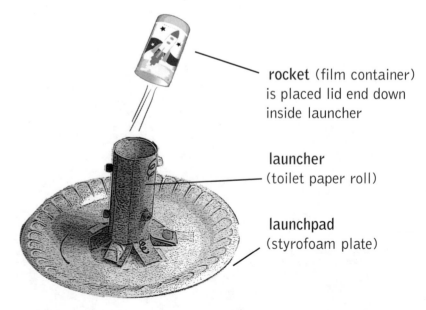

rocket (film container) is placed lid end down inside launcher

launcher (toilet paper roll)

launchpad (styrofoam plate)

72 Invisible Ink

Materials

- Round film container
- Adhesive label: **Invisible Ink**
- Lemon juice
- Pony bead
- Q-Tip

Directions

1. Fill the container about half full of lemon juice.

2. Hot-glue a pony bead in the center of the inside of the container lid.

3. Cut the Q-Tip in half and insert the cut end into the pony bead hole, securing it with more glue.

4. Give this new writing implement to a child and ask her to write some words with it by dipping the Q-Tip into the lemon juice and writing on a sheet of paper. Have the child hold up the paper to the light to reveal what she wrote.

Consider This:

Check to make sure the lid is securely attached to the container to prevent spills and to keep the lemon juice from evaporating quickly.

73 Window Garden

Materials

- Clear film container for each child
- Reproducible Label: _____'s Garden
- Potting soil
- Seeds (sunflower seeds and lima beans grow the quickest)

Directions

1. Fill each container half full of potting soil.

2. Add a seed, sprinkle a little soil on top, water lightly, and set in a sunny spot, watching daily for signs of growth.

3. Once the plant's roots are firmly established, have each child take her plant home and transplant it in the yard (if it is the right season) or in a larger container in the house.

Consider This:

a. Remind the children to remove the plant from the film container before transplanting it.

b. Have the children record their plants' growth on a weekly basis.

c. Use 2 types of seeds, such as a sunflower seed and a lima bean, and compare rates of growth.

d. Compare rates of growth using various locations in the classroom.

74 Search for the Stars

Materials

- Black film containers (with flat bottoms, not dimpled)
- Adhesive labels: **Search for the Stars, #_____**
- Constellations reproducible (see page 142)

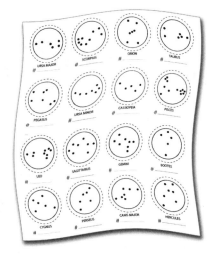

Directions

1. Make a copy of the reproducible, cut out a constellation, and temporarily tape it to the bottom of a film container.

2. Using a ballpoint pen, punch holes at the dots, piercing through both the paper and the container.

3. Repeat the procedure for the remaining constellations and containers.

4. Give a container to a child, along with a copy of the reproducible, and ask him to remove the lid and hold the container up to the light, looking through the end where the lid was attached.

5. Have the child record on his sheet the container number that matches the constellation.

Consider This:

a. Use gold stars, foam stars, or other materials for decorating the containers.

b. Ask parents or volunteers to help make the container constellations.

c. Pierce each child's initials into a container and label it _____'s Stars.

75 Mixing Colors

Materials

- 3 clear film containers
- 3 adhesive labels: **Red, Blue, Yellow**
- Red, blue, and yellow food coloring
- Paper towels
- 3 eyedroppers
- Mixing Colors reproducible (see page 143)

What color goes in the blanks?

red + yellow = orange

blue + red = purple

___ + ___ = green

Directions

1. Pour a little water into each container. Using a different eyedropper for each color, add a few drops of food coloring to the appropriate containers.

2. Let a child experiment by mixing drops of different-color water on a paper towel to see what secondary colors he can create from the 3 primary colors. Have the child record his results on the Mixing Colors reproducible.

Consider This:

You may want to tie an eyedropper to each container so the children won't use the same dropper for different colors.

76 Smartie Putty

Materials

- Film container of any kind
- Reproducible Label: ____'s **Smartie Putty**
- 2 teaspoons liquid starch (available at supermarkets and drugstores)
- A few drops food coloring
- 1 tablespoon Elmer's Glue-All (*not* Elmer's School Glue)
- Craft stick

Directions

1. Add the starch, food coloring, and glue to the film container in the order given, stirring with the craft stick after each addition. If the mixture is too sticky, add a little more starch.

2. Thoroughly combine the mixture, then turn it out into your hands and knead it to form a ball.

3. Show the class how the ball bounces and lifts print from a newspaper as you discuss how the liquids formed a solid when mixed together.

Consider This:

This also makes a fun gift for a child—and for a child to make and give to others.

77 Match the Sounds

Materials

- 12 black film containers
- Reproducible label: **Match the Sound**
- Rice, pasta, bells, small coins, rubber balls, and pencil erasers (or 6 items of your choice)
- Match the Sounds reproducible (see page 144)

Directions

1. Partially fill 2 film containers with one of the items. Repeat for the remaining items.

2. Number the lids of the containers 1–12, with 1–6 being the 6 different items and 7–12 being a repeat of these same 6 items, but not numbered in the same order as the first batch.

3. Have a child shake all the containers and match the sounds by drawing a line on the reproducible connecting the 2 numbers with the same sound. For example, if rice is in container 1 and container 10, the child would draw a line from number 1 to number 10 on the reproducible.

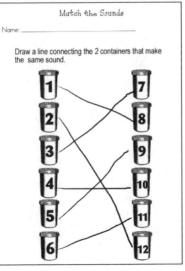

Match the Sounds

Name: _____

Draw a line connecting the 2 containers that make the same sound.

1 7
2 8
3 9
4 10
5 11
6 12

Consider This:

a. To keep the children from peeking inside the containers (and to prevent the lids from popping off), you might want to glue the lids on.

b. Change this to a "Match the Smells" activity by filling containers with items that have distinctive smells, such as flaked coconut, coffee beans, tea leaves, vinegar, vanilla extract, and peppermint extract. Proceed as above, except instead of having the children shake the containers, poke some holes through the lids and let the children sniff the contents. (If you are using liquids, put a few drops on a cotton ball and place it in the container.) Alter the reproducible so that it reads "Match the Smells" or create your own sheet. Again, you may want to glue the lids on to prevent spills and spying.

78 Kaleidoscope

Materials

- Clear film container for each child
- Sheets of clear plastic (such as sheet protectors)
- Mirrorlike paper (available at craft stores)
- Prism beads (available at craft stores)

Directions

1. For each child, place a container on a plastic sheet, trace around the base, and cut out the disk shape, being sure to cut just inside the traced line so the disk will fit inside the container.

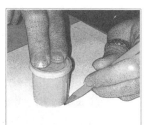

2. From the mirrorlike paper, cut a rectangular piece that measures about 2½" x 2" and fold it into thirds crosswise (along the shorter side). Tape it together to form a triangular-shape cylinder.

3. Use a hole punch to make a viewing hole in the container's lid.

4. To assemble the kaleidoscope, put the beads in the container, followed by the clear plastic disk and then the triangular cylinder, which may need to be trimmed slightly so that it does not extend above the top of the container. Snap on the lid.

5. As the children look through their kaleidoscopes while holding them up to the light and turning them, you can have a discussion on light reflection.

Consider This:

a. Increase or decrease the number of beads enclosed to see how this affects viewing.

b. Add glitter or gems to the bottom of the container.

79 Which Things Are Magnetic?

Materials

- Film container of any kind
- Adhesive label: **Which Things Are Magnetic?**
- Various small items, such as a dime, piece of aluminum foil, paper clip, rubber band, and brass fastener
- Which Things Are Magnetic? reproducible (see page 145)
- Small magnet

Directions

1. List all the small items on the reproducible before you photocopy it. Place the items in the container.

2. Give a child the container, a copy of the reproducible, and the magnet. Have her test each item with the magnet to see if it is magnetic, then record her observations on the reproducible.

Consider This:

To keep the magnet from disappearing, hot-glue it to the container lid or attach it to the container with string.

80 Can You Make Me Mix?

Materials

- Clear film container
- Cooking oil
- Food coloring

Directions

1. Fill the container one-third full of cooking oil. Add enough water so that the container is two-thirds full. Add a few drops of food coloring.

2. Securely fasten the lid to the container and let a child shake it and try to mix the contents. Let the container sit for 5–10 minutes and then have the children draw what they see in the container.

3. Have another child shake the container, let it sit for 5–10 minutes, and have the children draw what they see. Then repeat once more for a third time.

Consider This:

a. Have an older child write about why the contents will not mix.

b. To prevent a potential mess, glue the lid to the container.

81 Bug Cage

Materials

- Clear film container for each child
- Netting
- Grass
- Small live bug (such as a ladybug)

Directions

1. Using a knife, cut an "observation window" out of the side of each container and glue a piece of netting to the inside of the container over the window.

2. Put some grass in the bottom of each container and add the bug.

3. Have each child watch his bug for about 10 minutes, recording what it does.

4. Let the bugs go after the activity, but release them outside—not inside the classroom.

Consider This:

a. Have an older child write a story about the bug.
b. Make a small slit in the lid and insert both ends of a pipe cleaner into the slit to create a loop that can serve as a carrying handle.
c. Put a plastic bug inside instead of a real one and use it as a story starter.

82 Static

Materials

- Clear film container
- Confetti
- Glitter

Directions

1. Pour some confetti and glitter into the container and attach the lid.

2. Rub the container on various materials (hair, clothing, and so on) and have a child observe what happens to the contents of the container.

3. Have the child describe the experience orally or in writing, addressing questions such as the following: What causes the glitter and/or confetti to stick to the sides of the container? Are there times when the glitter sticks but the confetti does not? Why?

Consider This:

Rub the container in one direction on fake fur or colored cellophane. To receive a positive charge, you must rub it in one direction over and over again—not back and forth, which produces a positive and a negative charge.

83 Let It Snow

Materials

- 2 clear film containers
- Permanent marker

Directions

1. On a snowy day, take the containers outside and put snow in both—firmly packing the snow in one and loosely filling the other.

2. Have a child mark on each container what level she thinks the melted snow will reach.

3. Put the lids on the containers and let the contents melt naturally. Have another child mark the level of the melted snow on each container.

4. Compare the predicted levels to the actual levels and discuss this with the class.

Consider This:

If there is no snow, fill a container half full of water and mark the water level with the permanent marker. Securely attach the lid and place the container upright in the freezer. After the water is frozen, mark the level of the frozen water and compare it to the liquid level. Discuss why the levels are different.

84 Hibernation

Materials

- Film container of any kind for each child
- Reproducible Label: _____'s **Buddy, Do Not Disturb!**
- Art supplies to create animals (pom-poms, wiggly eyes, craft foam, and the like)

Directions

1. After a class discussion about animals that hibernate, have each child create one of the animals discussed.

2. Ask each child to predict the number of days it will be before her animal comes out of hibernation and to write that number on a slip of paper.

3. Mark each child's prediction on a graph. Have each child put the number and her animal in her film container and attach the lid.

4. Store all the containers in a shoe box and mark the days on a calendar when the animals will come out of hibernation.

5. Open the children's containers in chronological order (based on the children's predictions) and ask the class, "Do we think this animal would come out now?"

Consider This:

Have older children do research on their animals and write reports about them.

85 Flashlight

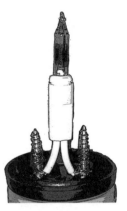

Materials

- Clear film container
- Black film container
- 9-volt battery
- 2 screws
- String of white (clear) Christmas lights

Directions

1. Cut off the bottom of the clear container. Place the battery inside the black container with the positive/negative end up.

2. Put the 2 lids together, top to top, and push the screws through both. Be sure that the heads of the screws line up with the positive/negative ends of the battery.

3. Cut out one of the lights from the string, leaving 2 inches of wire on both sides of the light.

4. Strip off a portion of the plastic covering from both wires and wrap one set of exposed wires around one screw end and the other set of

exposed wires around the other screw end.

5. Snap the lid with the screw heads onto the black container and the lid with the screw ends and light onto the clear container.

6. Turn the lids to turn the flashlight off and on. (When the screws touch the battery, the light comes on.)

Consider This:

a. Use a colored light for a different effect.
b. For safety purposes, do not cut off the bottom of the clear container for young children. Keeping the bottom on creates a light stick.

86 Bubbles

Materials

- Film container of any kind for each child
- Pipe cleaners
- 1 cup dishwashing liquid
- 2 cups warm water
- ¼ cup glycerin (available at drugstores)

Directions

1. Poke a small hole just large enough to push a pipe cleaner through the center of each lid.

2. Make blowing wands by cutting pipe cleaners into 3½" lengths. Push a pipe cleaner through the lid and bend one end (the end that will be inside the container) to form a loop, attaching it to the rest of the pipe cleaner. The straight part of the pipe cleaner protruding through the top of the lid serves as a handle.

3. Mix the dishwashing liquid, water, and glycerin together in a bottle or pitcher. Partially fill each container with the bubble solution. (This yields enough for 20–25 film containers.)

4. Take the children outside during the winter and give each a "bubble blower" and a container of solution. When they blow bubbles, the bubbles will freeze, and the children can catch them. Use this fun experiment for a science discussion about the bubbles.

Consider This:

a. Make 3 different solutions using 3 different dishwashing liquids and ask the children to predict which will produce the best bubbles (judged by the size and strength of the bubbles). Have them report back to you with their findings.

b. Make different solutions using different amounts of dishwashing liquid. Proceed as with the previous variation.

87 Evaporation

Materials

- 2 clear film containers (lids not needed)
- Permanent marker
- Water Evaporation reproducible (see page 146)

Directions

1. Fill both containers with an equal amount of water and mark the water level on each one with the permanent marker.

2. Place one of the containers in a sunny location and the other in a dark location, and mark the day's date on your classroom calendar.

3. Give each child a reproducible sheet and ask the children to predict how many days it will take for the water in each container to evaporate. Have them write their predictions on their sheet in the appropriate space.

4. Make a chart or graph showing each child's predictions.

5. Around the same time each day, have a different child check the containers. When a child sees that a container is empty, he should tell you right away.

6. Compare the children's predictions to the actual dates of evaporation and discuss the process of evaporation under sunny versus dark conditions.

Name: _____

Date experiment began _____

Sunny Location

Dark Location

Predicted number of days _____
Actual number of days _____

Predicted number of days _____
Actual number of days _____

Consider This:

Turn this into a math activity by comparing the chart or graph of the children's predictions to the actual dates of evaporation.

88 Moldy Cheese

Materials

- 3 clear film containers
- 3 adhesive labels: Moldy _____ Cheese
- 1 piece each of 3 types of cheese (such as cheddar, American, and Swiss)

Directions

1. Put a piece of cheese in each container, label the container with the kind of cheese inside, and attach the lid. Place the containers in a warm spot.

2. Have the children do daily observations and record the changes in the cheese—either by writing about them or by drawing pictures of what they see. Make sure they date each written observation or drawing and identify the type(s) of cheese.

Consider This:

a. Keep pieces of the cheese in separate containers in the refrigerator and, on the last day of the observation, compare the moldy cheese to the refrigerated cheese.
b. After the mold appears, put each piece of cheese under the microscope for the children to examine.
c. If the mold spreads to the container, discuss why this happens.

89 String Phones

Materials

- 2 film containers of any kind (black seems to carry sound best)
- 36" piece of yarn

Directions

1. Use a needlepoint needle to make a hole in the bottom of each container.

2. Wrap tape around both ends of the yarn to make threading it easier. Thread one end through the bottom of one container and the other end through the bottom of the other container. Knot each end so that the knot is inside the container.

3. Have 2 children each hold a container. While one child speaks into her container, the other holds his container to his ear. Since sound travels better over a taut connection, have them compare talking and listening with the yarn pulled taut and with it held loosely.

Consider This:

Vary the "connection" by using different types of cords, such as string or wire, to see if this changes how the sound is carried.

90 Parachute

Materials

- Film container of any kind for each child
- 3 pieces of thin wire, yarn, or string approximately 12" long, for each container
- Plastic grocery bags

Directions

1. Remove each container's lid and use a hole punch to make 3 evenly spaced holes around the top of the container. Position them so that the top of each hole is about ¼" from the top of the container.

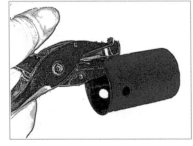

2. Thread 1 piece of wire (yarn or string) through one of the holes and either tie or twist the end to the container. Do the same with the other 2 pieces and other 2 holes.

3. Place a plate or another object that is 12" in diameter on a plastic bag, trace around the circumference, and cut out a 12" disk.

4. Punch 3 evenly spaced holes around the circumference of the circle about ½" from the outside edge. Insert the free end of one of the pieces of wire into a hole and tie (or twist) it securely to the plastic. Do the same with the other 2 pieces of wire and the other 2 holes. Now you have a plastic "parachute" attached to a film container, or "bucket."

5. Go outside and have the children go to different locations, such as the top of the slide, the sidewalk, and the middle of the playground, and have them toss their parachutes as high as they can. Then have them compare the rates of descent. Doing this on different days when the wind varies will provide an opportunity to discuss wind and lift.

Consider This:

a. Do this with a small plastic animal in the "bucket" to see what effect the additional weight might have.

b. Do a creative writing project based on what the animal being carried in the "bucket" might see.

91 Tooth Decay Experiment

Materials

- 3 clear film containers

- Bottled water

- Grape juice

- Cola

- Real teeth (child's baby teeth or teeth obtained from dentist)

- Tooth Decay reproducible (see page 147)

Directions

1. Fill each container three-quarters full of a different liquid and place a tooth in each one. Place the lids on the containers.

2. Have the children record in words or by drawing on a copy of the Tooth Decay reproducible what the teeth looked like on the first day of the experiment.

3. On Day 5, have the children make a second observation and record what is happening. (Use a spoon or tweezers to remove the teeth so the children can view them more easily.)

4. On Day 10, have them do a final observation and record what they see.

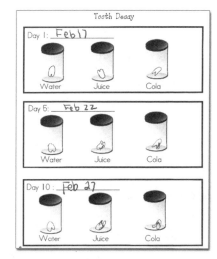

- -

Consider This:

a. Use milk in an additional container. (Remember, though, that it will spoil and curdle over time.)

b. Instead of teeth, use large pieces of white eggshell that have been cleaned with soap and water.

c. If you have a camera available, take photos every 2 or 3 days and create a bulletin board with the class, showing various stages of the experiment.

5. Art

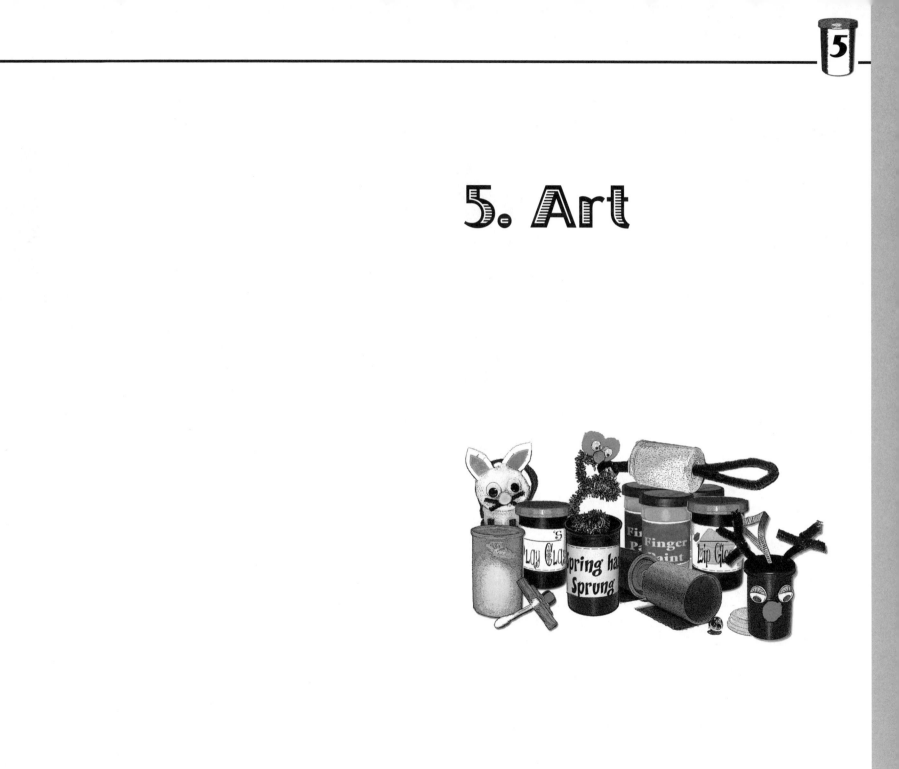

92 Spring Has Sprung

Materials

- Black film container for each child
- Reproducible label: **Spring Has Sprung**
- Strip of oak tag (approximately 1¾" x 18") for each child
- Pom-pom, small picture (of flower, bird, etc.), or small, lightweight item for each child

Directions

1. Have each child fold his strip of oak tag like an accordion and glue a pom-pom, picture, or other item to one end of it.

2. Have him glue the other end of the folded strip to the bottom of the container on the inside, then push the strip into the container and put on the lid.

3. Tell the child to remove the lid quickly and watch the pom-pom, picture, or other item spring out. If it does not, the item on the end might be too heavy. Have the child replace it and try again.

Consider This:

Add a face to the end of the strip and refer to it as a "Smartie in a Can."

93 Penguin

Materials

- Black film container with 2 lids
- Penguin Pattern Pieces reproducible (see page 148)
- Craft foam or construction paper
- Wiggly eyes

Directions

1. Using the reproducible, cut out the penguin pieces from the foam or construction paper.

2. Except for the hat, glue the penguin pieces and the wiggly eyes to the container.

3. Snap one lid onto the container, spread glue over the lid, and attach the hat brim to it. Glue the other lid to the brim so that the lid is upside down—the ridge side is facing up, and the smooth side is facing down.

4. Put a small amount of glue inside the ridge track and create a top hat by feeding the foam or paper strip into the ridge.

5. Fold down the top circle so it covers the hat and secure it in place by folding the tab and gluing it to the inside of the hat.

Consider This:

a. Instead of a penguin, decorate the container as a bunny, turkey, squirrel, snowman, or reindeer, using wiggly eyes, pipe cleaners, and other appropriate art supplies.

b. This makes a great "goodies" holder for small treats.

94 Play Clay Holder

Materials

- Film container of any kind for each child
- Reproducible label: _____ 's **Play Clay**
- 1 cup flour
- ½ cup salt
- 2 tablespoons vegetable oil
- 1 teaspoon alum (available at drugstores)
- Food coloring

Directions

1. Combine the flour, salt, oil, and alum in a small bowl, mixing well. Stir in a small amount of water at a time until the mixture has the consistency of bread dough. Add a drop or two of food coloring and knead until the color is well blended.

2. Fill each child's container with a portion of the play clay and snap on the lid.

3. Let children who finish projects earlier than others use this as an activity.

Consider This:

Make an edible play clay by combining 1 cup powdered milk, ½ cup peanut butter, and ¼ cup honey. This can be eaten but does not keep for more than a day. It yields enough dough for 3 children.

95 Rainbow Crayon

Materials

- Black film container with no inside lip
- Adhesive label: **Rainbow Crayon**
- Crayon stubs or pieces of different-color crayons

Directions

1. Chop or shave the crayons and put them in the film container, filling it one-half to three-quarters full.

2. Set the container in the microwave without the lid and heat on high for 2 minutes at a time, until you see a little bit of liquid forming on top. Do not allow the contents to liquefy completely, or the colors will blend into a brown blob.

3. Let the container cool, then roll it back and forth in your hands, until the mixture loosens from the sides of the container and can be moved without falling out.

4. Make a hole in the bottom of the container with a needlepoint needle and use a pen or pencil to push up the crayon for use. When finished, push it back down in the container and attach the lid for storage.

96 Totem Pole

Materials

- 4 film containers with lids
- Totem Pole Art reproducible (see page 149)
- Colored markers, pens, or pencils
- Sand

Directions

1. Give a child a copy of the Totem Pole Art reproducible and let her choose 4 that she wants to use. Have her color them and then glue each one to a container.

2. Have her then fill one of the containers with sand and place the lids on all the containers.

3. Tell her to glue the 4 containers together, stacked as a totem pole, but to place the one with the sand on the bottom to give the pole stability.

97 Lip Gloss

Materials

- Film container of any kind
- Reproducible label: _____ 's **Lip Gloss**
- Unflavored vegetable shortening, such as Crisco
- Powdered drink mix
- Craft stick

Directions

1. Fill a film container half full of shortening and microwave it on high for 30–45 seconds.

2. Add 1 teaspoon of the drink mix and blend thoroughly with a craft stick.

3. Refrigerate for 1 hour or until solid.

Consider This:

a. Keep extras on hand for children with chapped lips.
b. Give a child her own container labeled with her name.

98 Rain Stick

Materials

- 3 clear containers
- Free-flowing items that are small enough to flow easily through ½"-diameter holes (examples: rice, popcorn kernels, beads)

Directions

1. Make a ½"-diameter hole in the bottom of each container and in one of the lids. Be sure the holes line up when the containers are stacked.

2. Glue the containers together in this arrangement: bottom container is lid end down and hole end up; middle container has holes in the bottom and the lid; top container is bottom end down and lid end up.

3. Fill the top container with the small items and use the rain stick as a noisemaker or as a timer for transition times.

Consider This:

a. To extend the amount of time it takes for the items to flow through, glue on additional containers.

b. Another way to lengthen the flowing time and to change the sound is by putting a small piece of wire mesh inside one or more of the containers. Just be sure the openings in the mesh are large enough for the items to flow through.

99 Finger Paints

Materials

- Film containers of any kind
- Reproducible label: _____ **Finger Paint** (fill in the color)
- 3 tablespoons granulated sugar
- ½ cup cornstarch
- 2 cups cold water
- 4 different colors of food coloring
- Powdered laundry detergent

Directions

1. Combine the sugar and cornstarch in a small saucepan. Add the water and cook over low heat, stirring constantly, until the mixture is thoroughly blended.

2. Divide the mixture equally among 4 bowls and stir 3–5 drops of a different color of food coloring into each bowl.

3. Mix a pinch of detergent into each bowl, blending well. (The detergent helps with cleanup time!)

4. Divide the paints among the film containers, write the appropriate color on each label, and let the children use them as needed. (Makes enough for 20–25 film containers.)

Consider This:

Make sure the lids are securely attached when the finger paints are not in use.

100 Body Glitter

Materials

- Film containers of any kind
- Reproducible label: _____ 's Body Glitter
- ¼ cup aloe vera gel (available at drugstores)
- 1 teaspoon glycerin (available at drugstores)
- ¼ teaspoon fine polyester glitter of any color (available at craft stores)
- 5 drops any fragrance oil (optional)
- 1 drop food coloring (optional)

Directions

1. Combine the aloe vera gel and glycerin in a small bowl. Stir in the glitter and optional fragrance and color until well blended.

2. Divide the glitter among the film containers and attach the lids. (Makes enough for 2–3 film containers.)

Consider This:

Be sure to check for allergies before using fragrance oils.

101 Shakers

Materials

- Film container of any kind
- Any granulated material used in the classroom (examples: sand, glitter, salt)

Directions

1. Pierce small holes in the lid of the container with a ballpoint pen, needlepoint needle, or the pointed end of a mathematical compass.

2. Fill the container with the granulated material, attach the lid, and label. Store until needed.

Consider This:

a. For quick identification of the container's contents, glue some of the material onto the container.

b. Turn this into a gift by making two shakers and putting salt in one and pepper in the other. Label both and attach a gift tag (see page 151). You might also want to put a piece of tape over the holes to keep the contents from spilling.

102 Maraca

Materials

- Film container of any kind
- Craft stick
- Popcorn kernels and rice
- Craft foam, beads, sequins, ribbon, or other decorative items

Directions

1. Cut a slit in the lid or bottom of the container that is just wide enough for a craft stick to be inserted.

2. Push about ½" of the stick through the slit and glue it in place.

3. Loosely fill the container with popcorn kernels and rice and glue the lid to the container.

4. Give a child the decorative items, have her decorate the maraca as desired, and get ready to strike up the band!

103 Necklace

Materials

- Film container of any kind
- Piece of string, ribbon, or yarn approximately 24" long
- Pony bead

Directions

1. Use a needlepoint needle to make a hole in the lid of the container.

2. Thread both ends of the string through the hole in the lid, going from the outside to the inside.

3. Thread both ends through a pony bead and tie them together so they cannot slip through the bead. Secure the string with a drop of glue, if desired.

4. Give the necklace to a child so that she can carry home important objects, loose change, a lost tooth, and the like.

104 Bell

Materials

- Film container of any kind (lid not needed)
- Pipe cleaner
- Bell

Directions

1. Use a ballpoint pen or needlepoint needle to make a hole in the bottom of the container.

2. Thread both ends of the pipe cleaner through the hole in the bottom of the container and attach the bell so that it hangs inside.

3. Shake the container and sing "Jingle Bells"!

Consider This:

To make this even jazzier, decorate the container with stickers.

105 Rattlesnake

Materials

- 5 film containers of any kind
- String
- Pony beads (optional)
- Colored construction paper or craft foam
- Rice, popcorn kernels, or a small bell

Directions

1. Use a ballpoint pen or needlepoint needle to make a hole in the bottom of 1 container (this is the rattler's head), in the bottoms and lids of 3 containers (this is the rattler's body), and in the lid of 1 container (this is the rattler's tail). Make sure all the holes line up.

2. Use the string to tie the containers together in the order described in step 1, leaving some space between the containers either by tying knots or by using pony beads as spacers.

3. Decorate the rattlesnake with construction paper or craft foam. Add some rice, popcorn kernels, or a small bell to the "tail" container to create the rattler sound.

4. Attach a string to the rattler's head and use it as a pull toy.

106 Craft Foam Stamps

Materials

- Film container of any kind
- Lid for each stamp design
- Stamp Design Patterns reproducible (see page 150)
- Craft foam

Directions

1. Copy and cut out the reproducible patterns.
2. Trace the patterns onto craft foam and cut out the shapes.
3. Glue each shape to the top of a container lid.
4. Use the film container as the stamp handle. Each container can have multiple lids with various stamp designs.

Consider This:

See also the Stamp Handle activity on page 25.

107 Individual Ink Pad

Materials

- Film container of any kind
- Dense sponge (one without a lot of large holes)
- Ink
- Craft Foam Stamps (see left)

Directions

1. Cut the sponge into small pieces that will fit in the container. Fill the container to the top with these pieces, packing them down as you would brown sugar in a measuring cup.
2. Pour in some ink and attach the lid.
3. Use the pad with the Craft Foam Stamps.

108 Puffy Paint for Paper

Materials

- Film containers of any kind
- Equal parts salt, flour, and water
- Liquid tempera paint
- Q-Tips

Directions

1. Combine the salt, flour, and water in a small bowl until well blended. Stir some tempera paint into the mixture.

2. Pour the paint into the film containers and have the children use Q-Tips as brushes.

109 Mini Tool Chimes

Materials

- Film container of any kind
- Fishing line
- Small metal items such as screws, bells, nuts, and bolts
- Pipe cleaner

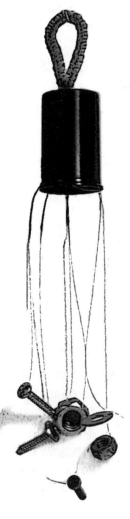

Directions

1. Use the pointed end of a mathematical compass to poke small holes around the top of the container, just below the lip.

2. Tie one end of a piece of fishing line to each of the metal items. Thread the other end of each line through a hole in the container and tie it off.

3. Use a ballpoint pen to make a hole in the bottom of the container. Make a loop with the pipe cleaner and thread both ends through the hole from the outside, twisting the ends together so they will not pull back out. This creates a hanger for the chime.

4. When sending the chime home with a child, push everything inside the container and attach the lid.

110 Pop-Out Sidewalk Chalk

Materials

- Film containers of any kind
- 1/3 cup quick-setting plaster of paris
- 1 teaspoon glitter
- 1 tablespoon powdered tempera paint
- 3 tablespoons water
- Craft stick

Directions

1. Combine all the ingredients in a paper cup by stirring with the craft stick until well blended.

2. Divide the mixture among the film containers and let sit for 30–45 minutes, or until hard. (Makes enough for 5–6 film containers.)

3. Use the containers as holders as well as storage for the chalk.

Consider This:

Dixie cups work well for mixing the ingredients.

111 Sand Art

Materials

- Clear film container
- Different-color craft sand
- Toothpick

Directions

1. Fill the container with layers of different-color sand. Poke the sand with the toothpick after each layer is added to create a design and to combine colors in patterns.

2. Glue the lid onto the container.

Consider This:

Make as a gift and attach a gift tag (see page 151).

112 Special Event/ Holiday Memories Ornament

Materials

- Film container of any kind for each child
- Decorative materials suitable for a particular holiday (examples: Halloween, Thanksgiving, Christmas, Hanukkah, Easter)
- Embroidery thread, metallic braid, or similar material
- Preprinted memories record sheet that you create (see step 3)

Directions

1. Give each child a film container. Have the children decorate the containers according to the event or holiday at hand, then write the date of the event on the bottom of the container.

2. Give the children embroidery thread, or some other suitable material, and have them make a loop and glue it to the container for hanging.

3. Give each child a copy of the memories sheet and have him fill it out, roll it up, and insert it in his container. The sheet should pertain to the actual event or holiday. For example, it might read:

Halloween, October 31, _____: My favorite candy was _____. My costume was _____. I went trick-or-treating with _____. I knocked on _____doors.

4. Have the children hang their containers in the classroom (on a Halloween goblin, Christmas tree, or Easter tree, for example). When appropriate, have each child read his sheet to the class, then put it back in his container and take it home.

Consider This:

a. Write the memories record sheet according to whether you have the children read them before or after the holiday.
b. Send extra copies of the sheet home with the children to be filled out every year.
c. To make a Vacation Memories Ornament, have the children collect sand, shells, pebbles, hemlock cones, and the like, and store them in their containers. Later, you can use them as a writing prompt or for a class discussion.

6. Gifts

· ·

Consider This:

Most of the following gift-making projects are simple and safe enough for young children to do on their own; however, be sure to read through the activity first before allowing a child to attempt it by himself, as there could be steps involved, such as using a knife or other pointed tool, that should be done by an adult or with adult supervision.

113 Pillbox

Materials

- Film container of any kind
- Reproducible label: _____'s Pillbox
- Crayons or colored markers or pens
- Gift Tag reproducible (see page 151)

Directions

1. Fill in the name of the person the gift is going to, and glue the label to the container.

2. Decorate, fill out, and attach the gift tag.

Consider This:

Use as a springboard activity for teaching children about the importance of never touching other people's medicines and vitamins.

114 Wish Jar

Materials

- Film container of any kind
- Reproducible label: Wish Jar
- My Wish for You Is . . . reproducible (see page 151)
- Gift Tag reproducible (see page 151)

Directions

1. Fill out a copy of the My Wish for You Is . . . reproducible with a message for the person the gift is for.

2. Roll up the reproducible, put it in the container, and attach the lid.

3. Decorate, fill out, and attach the gift tag.

Consider This:

This is a nice gift for Mother's Day, Father's Day, or someone's birthday.

115 Cup and Ball

Materials

- Round film container
- Paint stick or large craft stick
- Piece of string approximately 12" long
- Round wooden bead that is small enough to fit in container
- Gift Tag reproducible (see page 151)

Directions

1. Glue the container to one end of the paint or craft stick.
2. Notch both sides of the stick ½"–1" from the opposite end of the stick.
3. Tie one end of the string to the bead and the other end to the notched part of the stick. The game is played by holding the stick at the notched end, tossing the bead in the air, and trying to catch it in the container.

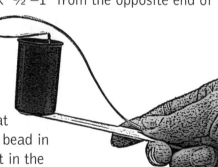

4. Decorate, fill out, and attach the gift tag.

Consider This:

For easy storage, keep the string attached to the stick, put the bead and the attached string inside the container, and snap on the lid.

116 Pincushion

Materials

- Round film container
- 3 cotton balls
- 4" square of fabric
- Rubber band
- Straight pins, needles, and safety pins
- Gift Tag reproducible (see page 151)

Directions

1. Put the lid on the film container, spread glue on the lid, and attach the cotton balls.
2. Place the fabric over the cotton balls so that it is centered. Put a rubber band around the container to hold the fabric in place.
3. Stick a straight pin in the cushion and put some more pins, needles, and safety pins inside the container.
4. Decorate, fill out, and attach the gift tag.

Consider This:

You may wish to include a sheet with instructions to add straight pins and needles at home for safety reasons.

117 Portable Aromatherapy

Materials

- Film container of any kind
- Reproducible label: **Portable Aromatherapy**
- 1–3 cotton balls
- Scented oil such as lavender, rose, or peppermint
- Gift Tag reproducible (see page 151)

Directions

1. Put the cotton ball(s) inside the container, add a few drops of the scented oil, and put the lid on the container.
2. Decorate, fill out, and attach the gift tag.

Consider This:

This is nice to use in the car or in a bureau drawer.

118 Cat Toy

Materials

- Film container of any kind
- Pony bead
- Piece of ribbon, string, or yarn
- Small beads or bells
- Gift Tag reproducible (see page 151)
- Catnip (optional)

Directions

1. Use a needlepoint needle to make a hole in the container lid. Thread the ribbon through the hole.
2. Tie a pony bead to the end of the ribbon on the inside of the lid.
3. Put a few beads or bells—or both—inside the container, along with some catnip, if desired.
4. Glue the lid to the container to prevent any spills and attach the label.

119 Travel Earring Holder

Materials

- Film container of any kind
- Craft foam
- Gift Tag reproducible (see page 151)

Directions

1. Cut a 1" x 1½" piece of craft foam.

2. Make pairs of tiny holes in the foam to attach pierced earrings.

3. Make 2 holes in the container lid for another pair of earrings.

4. Put the foam inside the container and snap on the lid.

5. Decorate the gift tag, fill it out so it reads "For Your Earrings," and attach it to the container.

Consider This:

Be sure to leave enough space between the holes so that there is room to insert the earrings.

120 Traveling Powder Container

Materials

- Film container with 2 lids that snap over top rim
- Reproducible label: **My Traveling Powder**
- Loose powder of some kind (examples: body, face, baby)
- Velcro
- Gift Tag reproducible (see page 151)

Directions

1. Use a needlepoint needle to make 5 or 6 holes in only one of the lids.

2. Fill the container with powder, snap on the lid with holes, and attach a gift tag that you have decorated and filled out.

3. Attach a small piece of Velcro to the top of the lid without holes and the matching piece of Velcro to the bottom of the film container. Use the lid without holes to keep the powder from spilling out during travel. Store this lid on the bottom of the container until needed.

121 Postage Stamp Dispenser

Materials

- Film container of any kind
- Reproducible label: **Stamp Dispenser**
- Colored markers or pens
- Gift Tag reproducible (see page 151)

Directions

1. Using a knife, cut a vertical slit in the side of the container, making sure it is wide enough for a stamp to pass through.

2. Glue the label to the container, being careful not to cover up the slit.

3. Decorate, fill out, and attach the gift tag.

123 Address Labels

Materials

- Film container of any kind
- Reproducible label: **Address Labels**
- Gift Tag reproducible (see page 151)

Directions

1. Create return address labels on the computer. (In Microsoft Word, go to Tools/Envelopes and Labels.)

2. Cut them apart into individual labels, put them inside the container, and snap on the lid.

3. Decorate, fill out, and attach the gift tag.

Consider This:

Use labels of any kind, including bookplates for children.

122 Travel Sunscreen

Materials

- Film container of any kind
- Reproducible label: _____'s Sunscreen
- Sunscreen (optional)
- Gift Tag reproducible (see page 151)

Directions

1. Fill the container with sunscreen, if using, and attach the lid.

2. Fill in the name of the person the gift is going to and glue on the label.

3. Decorate, fill out, and attach the gift tag.

124 Office Supplies

Materials

- 3 film containers of any kind
- 3 reproducible labels: **Paper Clips, Rubber Bands, Tacks**
- Paint stick or large craft stick
- Gift Tag reproducible (see page 151)

Directions

1. Glue the containers onto the paint or craft stick, spacing them evenly and making sure all the labels face in the same direction.

2. Decorate the gift tag, fill it out so it reads "Office Supplies," and attach it to the stick.

Consider This:

For more stability, glue half a craft stick to each end of the paint (or large) stick.

125 Note Holder

Materials

- Film container of any kind (lid not needed)
- Reproducible label: **I'll Hold It for You!**
- Plaster of paris
- Large paper clip
- Gift Tag reproducible (see page 151)

Directions

1. Fill the container with plaster of paris.

2. When it is almost hardened, stick one end of the paper clip in it so that the clip is standing vertically.

3. Set the container aside until the plaster has hardened completely.

4. Decorate, fill out, and attach the gift tag.

Consider This:

Cover the container with a photocopy of a picture of the child who is giving the gift.

126 Hair Kit

Materials

- Film container of any kind
- Reproducible label: **Hair Kit**
- Tiny hair clips, rubber bands, and bobby pins
- Gift Tag reproducible (see page 151)

Directions

1. Decorate the container lid by gluing on a tiny hair clip.
2. Put the clips, rubber bands, and bobby pins inside the film container and snap on the lid.
3. Decorate, fill out, and attach the gift tag.

127 Friendship Bracelet Kit

Materials

- Film container of any kind
- Reproducible label: **Friends Forever**
- Colored beads and/or letter beads
- Piece of elastic or plastic cord approximately 12" long
- Gift Tag reproducible (see page 151)

Directions

1. Select the beads desired.
2. Enclose them, along with the elastic or cord, in the container. (The gift is meant to be made by the person who receives it.)
3. Snap on the lid.
4. Decorate, fill out, and attach the gift tag.

128 Refrigerator Magnet

Materials

- Film container of any kind (lid not needed)
- Gift Tag reproducible (see page 151)
- Green florist foam (available at flower shops, nurseries, and craft stores)
- Pipe cleaners
- Small beads
- Wire
- Self-sticking magnet

Directions

1. Use a ballpoint pen or needlepoint needle to make a hole on opposite sides of the container right below the top.

2. Thread one end of a pipe cleaner through one hole and the other end through the other hole, twisting the ends so they will not slip through. This creates a handle for the container.

3. Insert the florist foam until it is almost up to the top of the container but below the holes.

4. Make flowers with pieces of pipe cleaner, beads, and wire. Stick the flower stems into the foam. Attach the magnet to the back of the container.

5. Fill out the gift tag so it reads "For Holding Up My Special School Papers" or "Look What I Did!"

129 Kissing Frog

Materials

- Film container of any kind
- Reproducible label: **I've been kissed by a frog!**
- Construction paper, pom-poms, pipe cleaners, and craft foam
- 2 Hershey's Kisses
- Gift Tag reproducible (see page 151)

Directions

1. Make the parts of a frog's body from the construction paper, pom-poms, pipe cleaners, and foam, being sure the frog has big lips.

2. Glue the frog parts to the film container.

3. Put the Hershey's Kisses in the container and snap on the lid. Fill out a gift tag that reads: "_____ Gets a Kiss from a Frog! From _____."

130 Toothbrush Holder

Materials

- Film container with a flat lid
- Reproducible label: _____'s Toothbrush Holder
- Gift Tag reproducible (see page 151)

Directions

1. Cut an opening in the lid that is large enough for the handle on a toothbrush to be pushed through.
2. Decorate, fill out, and attach a gift tag to the container.

Consider This:

Present this with a new toothbrush enclosed.

131 Extra Stuff for Fishing

Materials

- Film container of any kind
- Reproducible label: **Fishing Stuff**
- Gift Tag reproducible (see page 151)

Directions

1. Glue the label to the film container and cover it completely with clear tape.
2. Fill out the gift tag so it reads "Fishing Tackle."

Consider This:

Make a waterproof money holder in the same way and attach a gift tag that reads "Keep Your Money Dry While You Fish."

Reproducibles

Homework Pass (pages 16 & 20)

Homework PASS

Homework PASS

Homework PASS

Homework PASS

Homework PASS

Homework PASS

Homework PASS

Homework PASS

Homework PASS

Homework PASS

Take your time.

I believe in you!

I'm here if you need me.

Do your best!

I know you can.

Please try.

Don't give up!

You are a smartie!

You can do it.

Sports	Fruits	Animals
baseball	banana	elephant
football	apple	horse
tennis	orange	rabbit
basketball	grapes	cow
hockey	watermelon	turkey
soccer	pear	bee

Sight Words List (page 30)

Week #1	Week #2	Week #3	Week #4
the	in	he	as
of	is	was	with
and	you	for	his
a	that	on	they
to	it	are	I

Week #5	Week #6	Week #7	Week #8
at	or	but	we
be	one	not	when
this	had	what	your
have	by	all	can
from	word	were	said

Excerpted from *The Reading Teacher's Book of Lists* by Edward Bernard Fry, Jacqueline E. Kress, and Dona Fountoukidis. © 2000 Jossey-Bass.
Reprinted by permission of John Wiley & Sons, Inc.

Sight Words List (page 30)

Week #9	Week #10	Week #11	Week #12
there	she	will	many
use	do	up	then
an	how	other	them
each	their	about	these
which	if	out	so

Week #13	Week #14	Week #15	Week #16
some	him	two	number
her	into	more	no
would	time	write	way
make	has	go	could
like	look	see	people

106

Excerpted from *The Reading Teacher's Book of Lists* by Edward Bernard Fry, Jacqueline E. Kress, and Dona Fountoukidis. © 2000 Jossey-Bass. Reprinted by permission of John Wiley & Sons, Inc.

Sight Words List (page 30)

Week #17	Week #18	Week #19	Week #20
my	call	long	come
than	who	down	made
first	oil	day	may
water	now	did	part
been	find	get	over

Week #21	Week #22	Week #23	Week #24
new	work	me	after
sound	know	back	thing
take	place	give	our
only	year	most	just
little	live	very	name

Excerpted from *The Reading Teacher's Book of Lists* by Edward Bernard Fry, Jacqueline E. Kress, and Dona Fountoukidis. © 2000 Jossey-Bass. Reprinted by permission of John Wiley & Sons, Inc.

Color and Shape Words (page 31)

black	blue	brown	green	orange	pink	red	yellow
circle	crescent	diamond	oval	rectangle	square	star	triangle

Word Families (page 31)

at	et	it	op	ug
an	en	in	ot	un
ap	ed	ip	og	up
ad	er	id	on	ub
all	ell	ill	oll	ull
ash	eg	ick	ob	um

Alphabet I (pages 31 & 38)

a	e	i	o	u
b	c	d	f	g
h	j	k	l	m
n	p	q	r	s
t	v	w	x	y
z				

Name _____

Real Words	Nonsense Words

a b c d e

f g h i j

k l m n o

p q r s t

u v w x y

z

1st Guess ☐

2nd Guess ☐

3rd Guess ☐

4th Guess ☐

I am between _____ and _____ in the alphabet.

I am found in the word _____.

I make this sound:

_____.

1st Guess ☐

2nd Guess ☐

3rd Guess ☐

4th Guess ☐

I am between _____ and _____ in the alphabet.

I am found in the word _____.

I make this sound:

_____.

1st Guess ☐

2nd Guess ☐

3rd Guess ☐

4th Guess ☐

I am between _____ and _____ in the alphabet.

I am found in the word _____.

I make this sound:

_____.

Setting, Action, and Characters (page 33)
Setting (Where?)

baseball field	pool	woods	football game	
circus	store	zoo	mall	
movies	soccer	school	restaurant	
beach	moon	jungle	desert	
island	rain forest	amusement park	picnic	

Action (What?)

looks	learns	visits	sees
faints	falls down	eats	gets lost
digs	checks	reads	sleeps
drinks	climbs	swims	grows
captures	smells	hears	finds

Setting, Action, and Characters (page 33)

Characters (Who?)

raccoon	ferret	bat	butterfly					
bird	donkey	lizard	pig					
ladybug	spider	duck	tiger					
monkey	mouse	cat	frog					
lion	ostrich	cow	turtle					

Contractions (page 34)

I am	he is	had not
they are	it is	has not
we are	she is	have not
you are	that is	is not
he has	there is	must not
it has	what is	should not
she has	where is	was not
what has	who is	were not
where has	are not	will not
who has	cannot	would not
I have	could not	let us
they have	did not	he will
we have	do not	I will
you have	does not	she will

Contractions (page 34)

they will	I'm	he's
we will	they're	it's
you will	we're	she's
he would	you're	that's
I would	he's	there's
she would	it's	what's
they would	she's	where's
we would	what's	who's
who would	where's	aren't
you would	who's	can't
	I've	couldn't
	they've	didn't
	we've	don't
	you've	doesn't

she'll		hadn't
they'll		hasn't
we'll		haven't
you'll		isn't
he'd		mustn't
I'd		shouldn't
she'd		wasn't
they'd		weren't
we'd		won't
who'd		wouldn't
you'd		let's
		he'll
		I'll

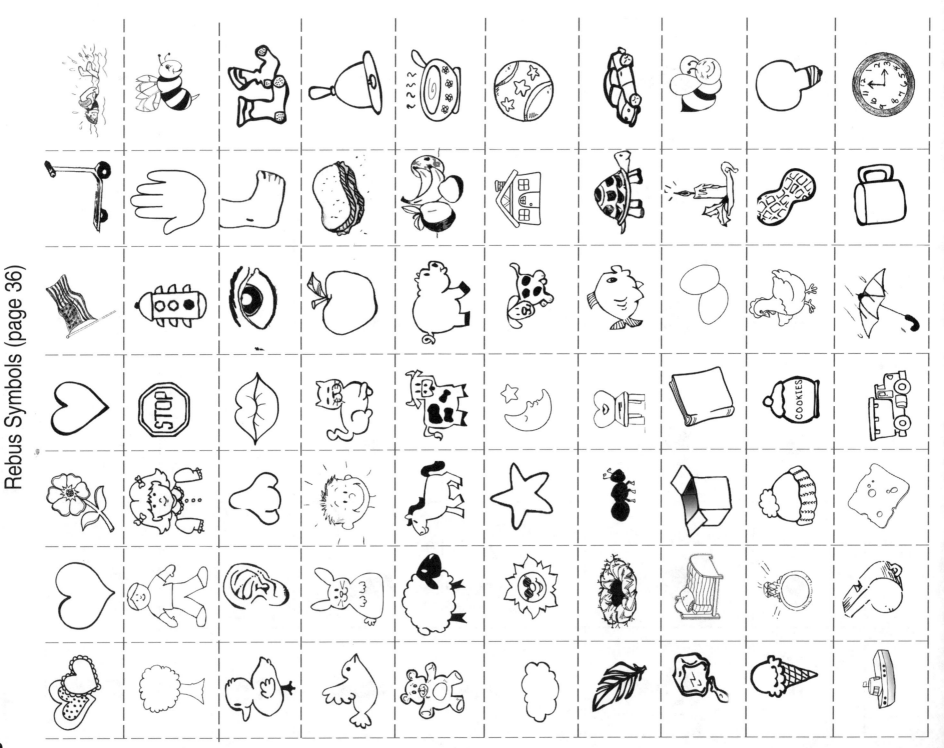

Reading Robot (page 36)

Title _____

Date Completed _____

Book Number # _____

Title _____

Date Completed _____

Book Number # _____

Title _____

Date Completed _____

Book Number # _____

Title _____

Date Completed _____

Book Number # _____

Title _____

Date Completed _____

Book Number # _____

Word Play (page 40)

Name _____

2-Letter Words	3-Letter Words	4-Letter Words

**Open House
Scavenger Hunt**

Find these items in
our classroom:

_____
_____
_____
_____

**Open House
Scavenger Hunt**

Find these items in
our classroom:

_____
_____
_____
_____

**Open House
Scavenger Hunt**

Find these items in
our classroom:

_____
_____
_____
_____

**Open House
Scavenger Hunt**

Find these items in
our classroom:

_____
_____
_____
_____

**Open House
Scavenger Hunt**

Find these items in
our classroom:

_____
_____
_____
_____

Message in a Container (page 42)

If you find me, please write how I spent my day and return me to _____'s classroom.
_____
_____
_____

If you find me, please write how I spent my day and return me to _____'s classroom.
_____
_____
_____

If you find me, please write how I spent my day and return me to _____'s classroom.
_____
_____
_____

Alphabet III (page 44)

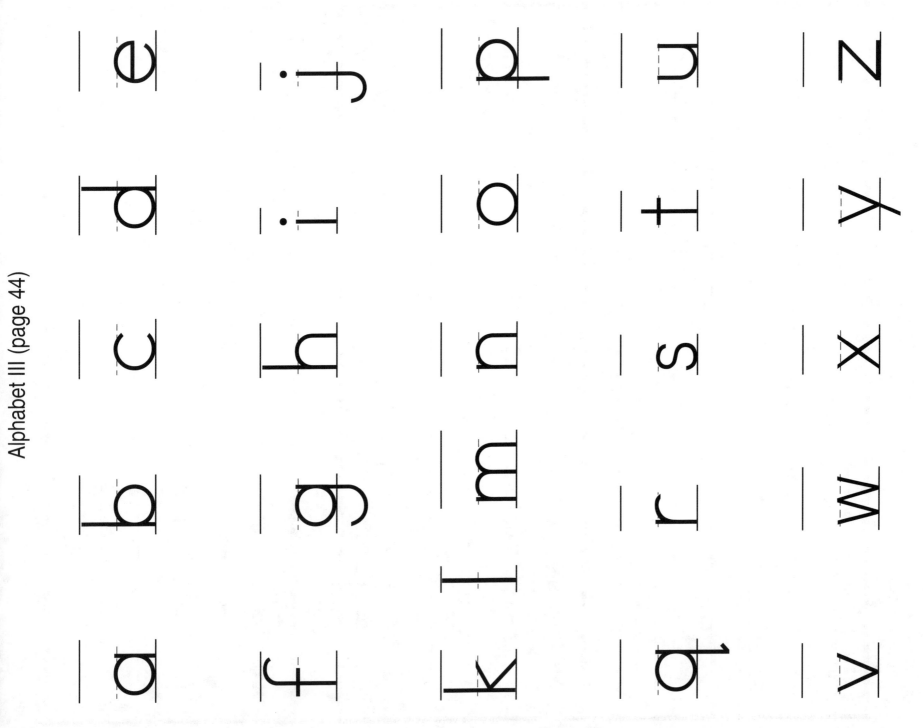

Ask Me About . . . (page 45)

Card 1
Name _____

Ask about _____.

_____

_____

_____

Please write your child's response on the lines provided and return it in the container to school.

Card 2
Name _____

Ask about _____.

_____

_____

_____

Please write your child's response on the lines provided and return it in the container to school.

Card 3
Name _____

Ask about _____.

_____

_____

_____

Please write your child's response on the lines provided and return it in the container to school.

Card 4
Name _____

Ask about _____.

_____

_____

_____

Please write your child's response on the lines provided and return it in the container to school.

Card 5
Name _____

Ask about _____.

_____

_____

_____

Please write your child's response on the lines provided and return it in the container to school.

127

What if the world was flat?

What if the sky was black?

What if the sun never shined?

What if cows had 6 legs?

What if you attended school only on holidays?

What if there were 8 days in a week?

What if kids had to do the grocery shopping?

What if there were no cars?

What if there was only one TV station?

What if

What if

Book Review

Reviewer's name: _____
Title: _____
Author: _____
Why I like or dislike this book: _____

Book Review

Reviewer's name: _____
Title: _____
Author: _____
Why I like or dislike this book: _____

Book Review

Reviewer's name: _____
Title: _____
Author: _____
Why I like or dislike this book: _____

Book Review

Reviewer's name: _____
Title: _____
Author: _____
Why I like or dislike this book: _____

Book Review

Reviewer's name: _____
Title: _____
Author: _____
Why I like or dislike this book: _____

Book Review (page 47)

1-, 2- and 3-Syllable Words (page 48)

applesauce	baby	ant
celebrate	bottle	book
difficult	brother	cold
excellent	candy	desk
family	furry	fun
Halloween	garden	game
hamburger	happy	hold
holiday	little	love
library	music	meet
mockingbird	oatmeal	nine
potato	party	off
seventeen	rabbit	run
Thanksgiving	student	sky
yesterday	yellow	train

0	1	2	3
4	5	6	7
8	9	10	zero
one	two	three	four
five	six	seven	eight
nine	ten		

(page 51)

Piggy in the Mud

Side	Back	Fours	Snout

Piggy
Master

_____
(name)

Piggy
Points

TOTALS

133

My Time Line

I was born   Ist tooth   walked   first word   sibling was born   started school

My Time Line

I was born   Ist tooth   walked   first word   sibling was born   started school

_____ Time Line

_____ Time Line

Dice Probability (page 55)

When you roll a die, it can land on: 1, 2, 3, 4, 5, or 6.

Predict which number you think it will land on most often.

_____

Why do you think this? _____

Roll the die _____ times and
record your results in the chart below.

1	
2	
3	
4	
5	
6	

How many times did it land on:
1 ___ 2 ___ 3 ___ 4 ___ 5 ___ 6 ___

1	2	3	4	5	6	7	8	9	10
11	12	13	14	15	16	17	18	19	20
21	22	23	24	25	26	27	28	29	30
31	32	33	34	35	36	37	38	39	40
41	42	43	44	45	46	47	48	49	50
51	52	53	54	55	56	57	58	59	60
61	62	63	64	65	66	67	68	69	70
71	72	73	74	75	76	77	78	79	80
81	82	83	84	85	86	87	88	89	90
91	92	93	94	95	96	97	98	99	100

Fact Family Game (page 57)

Fact Family

Left Hand	Right Hand

Weigh Me! (page 59)

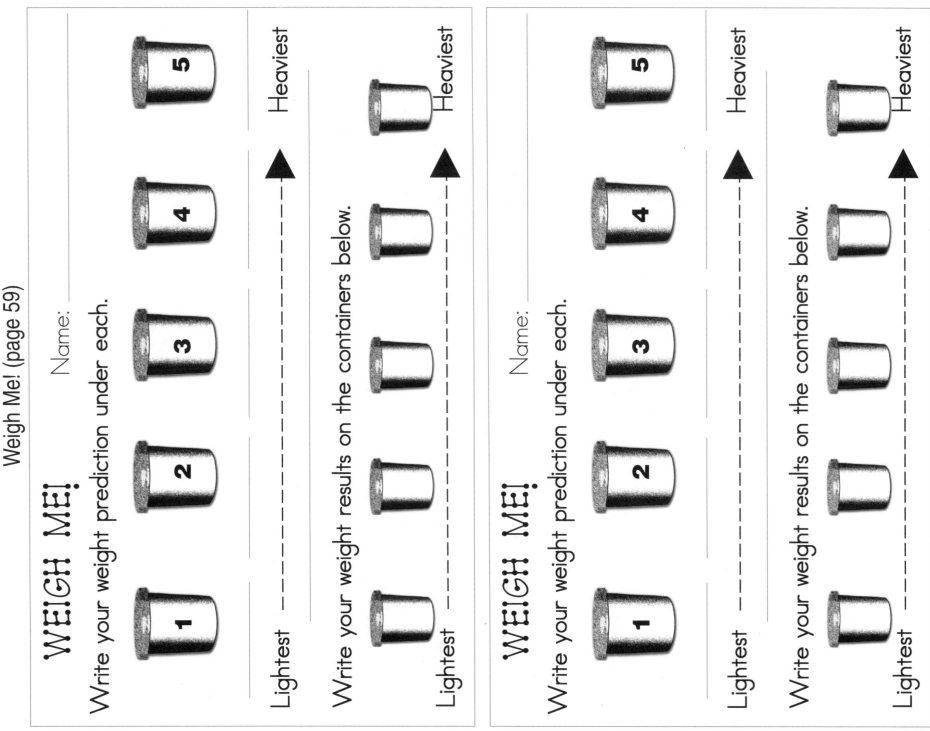

WEIGH ME!

Name: _____

Write your weight prediction under each.

5 4 3 2 1

Lightest - - - - - - ▲ Heaviest

Write your weight results on the containers below.

Lightest - - - - - - ▲ Heaviest

WEIGH ME!

Name: _____

Write your weight prediction under each.

5 4 3 2 1

Lightest - - - - - - ▲ Heaviest

Write your weight results on the containers below.

Lightest - - - - - - ▲ Heaviest

139

Penny Toss (page 61)

If you toss a penny in the air, it will land either heads or tails up. Which way do you think it will land most often?

_____

Why do you think this?

_____

_____

Toss the penny _____ times and record your results in the chart below.

Heads	
Tails	

How many times did it land heads up? _____
How many times did it land tails up? _____

Cross It Out

Player 1: _____

Player 2: _____

1		7
2		8
3		9
4	0	10
5		11
6		12

Cross It Out

Player 1: _____

Player 2: _____

1		7
2		8
3		9
4	0	10
5		11
6		12

Cross It Out

Player 1: _____

Player 2: _____

1		7
2		8
3		9
4	0	10
5		11
6		12

Constellations (page 66)

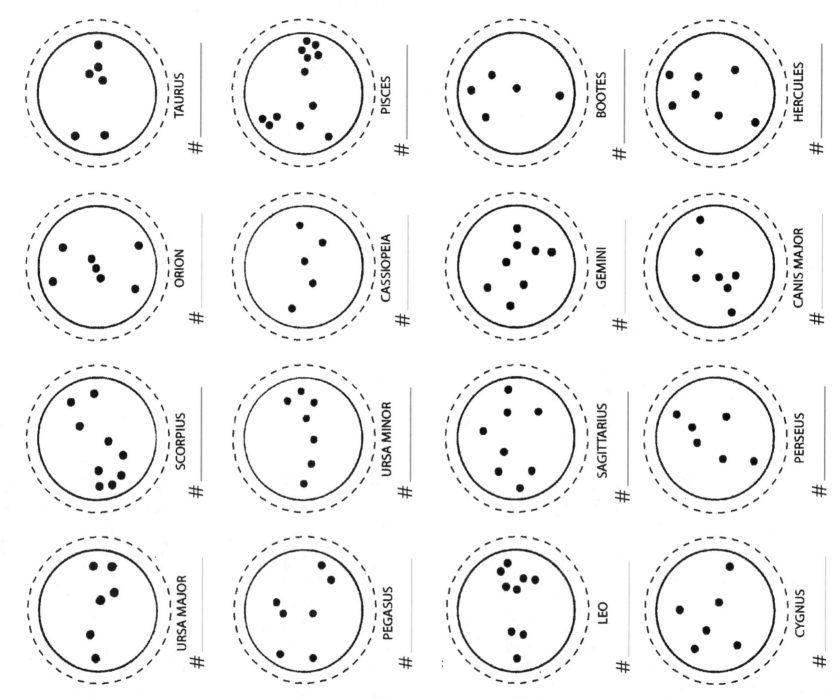

TAURUS

\# _____

PISCES

\# _____

BOOTES

\# _____

HERCULES

\# _____

ORION

\# _____

CASSIOPEIA

\# _____

GEMINI

\# _____

CANIS MAJOR

\# _____

SCORPIUS

\# _____

URSA MINOR

\# _____

SAGITTARIUS

\# _____

PERSEUS

\# _____

URSA MAJOR

\# _____

PEGASUS

\# _____

LEO

\# _____

CYGNUS

\# _____

142

Name: _____

What color goes in the blanks?

_____ + _____ = orange

_____ + _____ = purple

_____ + _____ = green

Mixing Colors

Name: _____

What color goes in the blanks?

_____ + _____ = orange

_____ + _____ = purple

_____ + _____ = green

Name: _____

Match the Sounds (page 68)

Draw a line connecting the 2 containers that make the same sound.

Name: _____

Which Things Are Magnetic? (page 70)

Item	YES	NO

Water Evaporation (page 75)

Name: _____

Date experiment began _____

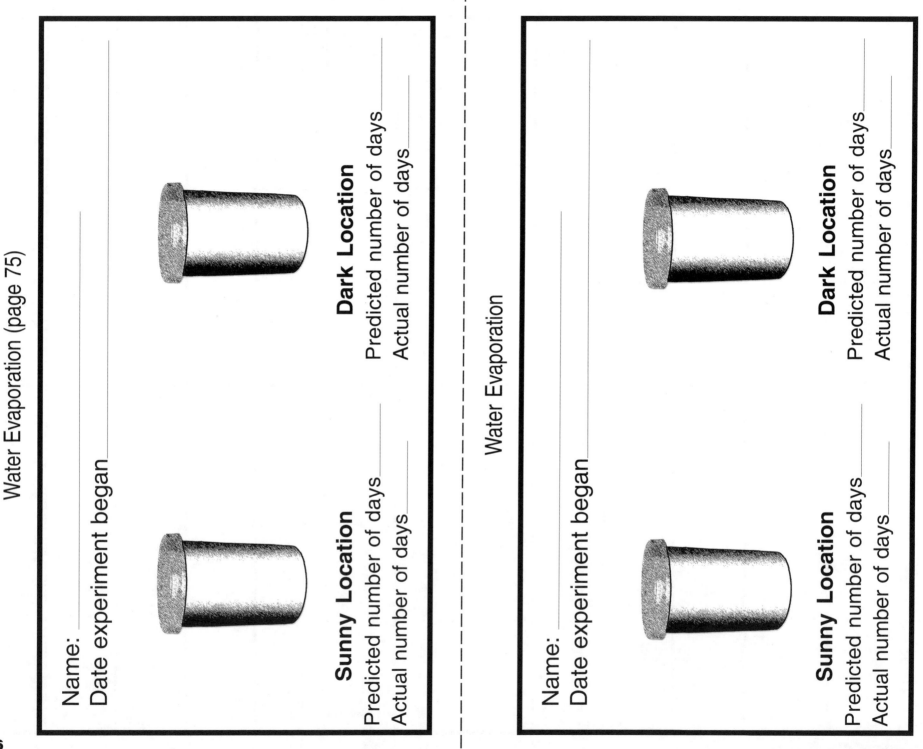

Sunny Location

Predicted number of days _____

Actual number of days _____

Dark Location

Predicted number of days _____

Actual number of days _____

Water Evaporation

Name: _____

Date experiment began _____

Sunny Location

Predicted number of days _____

Actual number of days _____

Dark Location

Predicted number of days _____

Actual number of days _____

Tooth Decay (page 78)

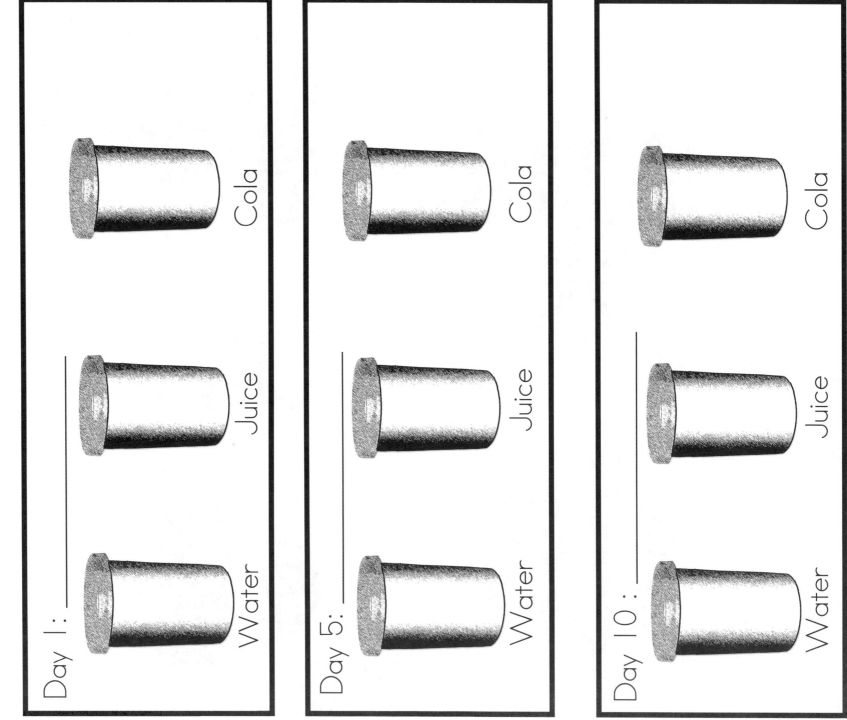

Day 1: _____   Water   Juice   Cola

Day 5: _____   Water   Juice   Cola

Day 10: _____   Water   Juice   Cola

Penguin Pattern Pieces (page 80)

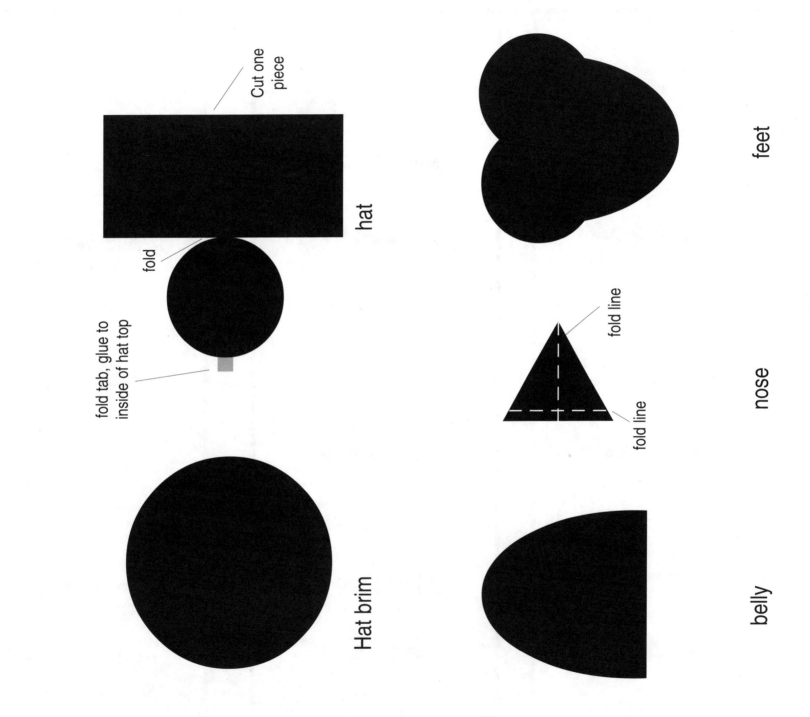

Cut one piece

hat

fold

fold tab, glue to
inside of hat top

Hat brim

feet

fold line

fold line

nose

belly

My Wish for You Is . . . . . (page 92)

My Wish for you is _____

My Wish for you is _____

Gift Tag

TO:

FROM:

TO:

FROM:

TO:

FROM:

Index to Text

Blank Grid (pages 37 & 48)

122

INDEX to Reproducible Labels

(To make these adhesive, we recommend photocopying them on #5260 Avery labels or something similar.)

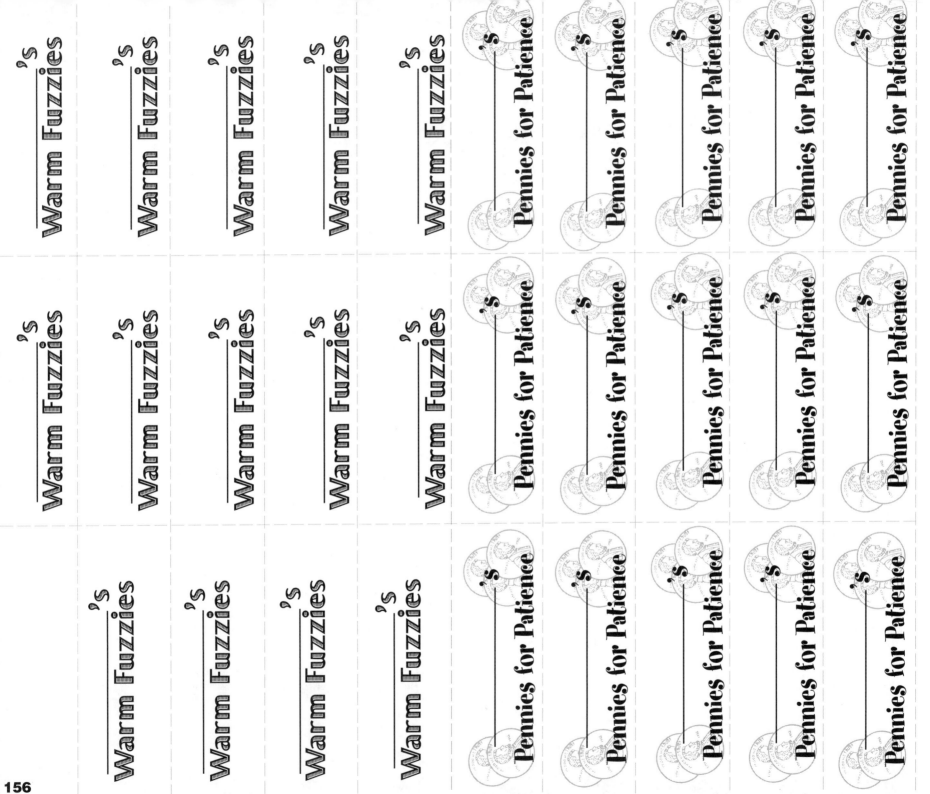

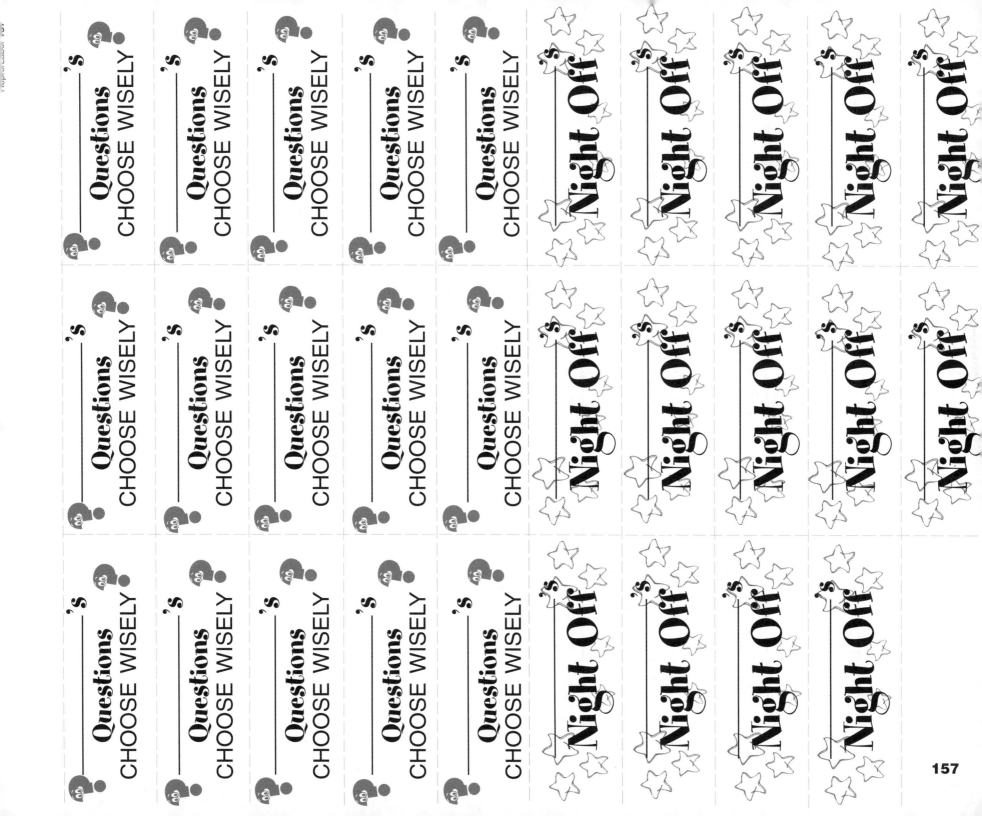

_____'s
Questions
CHOOSE WISELY

_____'s
Questions
CHOOSE WISELY

_____'s
Questions
CHOOSE WISELY

_____'s
Questions
CHOOSE WISELY

_____'s
Questions
CHOOSE WISELY

_____'s
Questions
CHOOSE WISELY

_____'s
Questions
CHOOSE WISELY

_____'s
Questions
CHOOSE WISELY

_____'s
Questions
CHOOSE WISELY

_____'s
Questions
CHOOSE WISELY

_____'s
Questions
CHOOSE WISELY

_____'s
Questions
CHOOSE WISELY

_____'s
Questions
CHOOSE WISELY

_____'s
Questions
CHOOSE WISELY

_____'s
Night Off

_____'s
Night Off

_____'s
Night Off

_____'s
Night Off

_____'s
Night Off

_____'s
Night Off

_____'s
Night Off

_____'s
Night Off

_____'s
Night Off

_____'s
Night Off

_____'s
Night Off

_____'s
Night Off

_____'s
Night Off

_____'s
Night Off

157

___'s

CENTERS
Look What I Finished!

___'s

CENTERS
Look What I Finished!

___'s

CENTERS
Look What I Finished!

___'s

CENTERS
Look What I Finished!

___'s

CENTERS
Look What I Finished!

___'s

CENTERS
Look What I Finished!

___'s

CENTERS
Look What I Finished!

___'s

CENTERS
Look What I Finished!

___'s

CENTERS
Look What I Finished!

___'s

CENTERS
Look What I Finished!

___'s

CENTERS
Look What I Finished!

___'s

CENTERS
Look What I Finished!

___'s

CENTERS
Look What I Finished!

___'s

CENTERS
Look What I Finished!

___'s

Back-to-School Kit

___'s

Back-to-School Kit

___'s

Back-to-School Kit

___'s

Back-to-School Kit

___'s

Back-to-School Kit

___'s

Back-to-School Kit

___'s

Back-to-School Kit

___'s

Back-to-School Kit

___'s

Back-to-School Kit

___'s

Back-to-School Kit

___'s

Back-to-School Kit

___'s

Back-to-School Kit

___'s

Back-to-School Kit

___'s

Back-to-School Kit

___'s

Back-to-School Kit

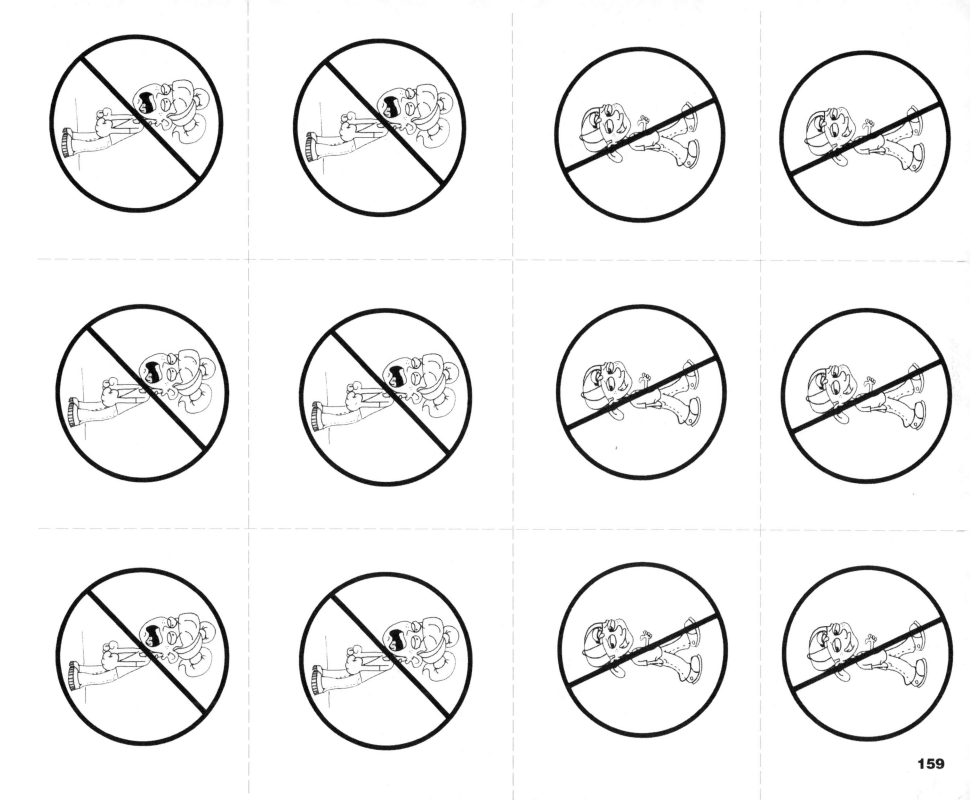

Tooth Suitcase

's

Tooth Suitcase

's

Tooth Suitcase

's

Tooth Suitcase

's

Tooth Suitcase

's

Project Pieces for

Project Pieces for

Project Pieces for

Project Pieces for

Project Pieces for

Tooth Suitcase

's

Tooth Suitcase

's

Tooth Suitcase

's

Tooth Suitcase

's

Tooth Suitcase

's

Project Pieces for

Project Pieces for

Project Pieces for

Project Pieces for

Project Pieces for

Tooth Suitcase

's

Tooth Suitcase

's

Tooth Suitcase

's

Tooth Suitcase

's

Project Pieces for

Project Pieces for

Project Pieces for

Project Pieces for

Project Pieces for

's
Counting Container

's
Counting Container

's
Counting Container

's
Counting Container

's
Counting Container

's
Counting Container

's
Counting Container

's
Counting Container

's
Counting Container

's
Counting Container

's
Counting Container

's
Counting Container

's
Counting Container

's
Counting Container

's
Counting Container

Sight Words
Week # _____

Sight Words
Week # _____

Sight Words
Week # _____

Sight Words
Week # _____

Sight Words
Week # _____

Sight Words
Week # _____

Sight Words
Week # _____

Sight Words
Week # _____

Sight Words
Week # _____

Sight Words
Week # _____

Sight Words
Week # _____

Sight Words
Week # _____

Sight Words
Week # _____

Sight Words
Week # _____

Sight Words
Week # _____

Word Family _____

Word Family _____

Word Family _____

Word Family _____

Word Family _____

Word Family _____

Word Family _____

Word Family _____

Word Family _____

Word Family _____

Word Family _____

Word Family _____

Word Family _____

Word Family _____

Word Family _____

_____'s Pet Rock

_____'s Pet Rock

_____'s Pet Rock

_____'s Pet Rock

_____'s Pet Rock

_____'s Pet Rock

_____'s Pet Rock

_____'s Pet Rock

_____'s Pet Rock

_____'s Pet Rock

_____'s Pet Rock

_____'s Pet Rock

_____'s Pet Rock

_____'s Pet Rock

_____'s Pet Rock

Rebus Story

Rebus Story

Rebus Story

Rebus Story

Rebus Story

Rebus Story

Rebus Story

Rebus Story

Rebus Story

Rebus Story

Rebus Story

Rebus Story

-'s Reading ROBOT

-'s Reading ROBOT

-'s Reading ROBOT

-'s Reading ROBOT

-'s Reading ROBOT

-'s Reading ROBOT

-'s Reading ROBOT

-'s Reading ROBOT

-'s Reading ROBOT

-'s Reading ROBOT

-'s Reading ROBOT

-'s Reading ROBOT

Rhymes with

Rhymes with

Rhymes with

Words

Words

_____'s Scavenger Hunt

_____'s Scavenger Hunt

_____'s Scavenger Hunt

All About _____

All About _____

All About _____

All About _____

_____'s Mini Mementos

_____'s Mini Mementos

_____'s Mini Mementos

_____'s Scavenger Hunt

_____'s Scavenger Hunt

_____'s Scavenger Hunt

All About _____

All About _____

All About _____

All About _____

_____'s Mini Mementos

_____'s Mini Mementos

_____'s Mini Mementos

_____'s Scavenger Hunt

_____'s Scavenger Hunt

All About _____

All About _____

All About _____

All About _____

_____'s Mini Mementos

_____'s Mini Mementos

_____'s Mini Mementos

What am I?
Hint #1 _____
Hint #2 _____
Hint #3 _____

What am I?
Hint #1 _____
Hint #2 _____
Hint #3 _____

What am I?
Hint #1 _____
Hint #2 _____
Hint #3 _____

What am I?
Hint #1 _____
Hint #2 _____
Hint #3 _____

What am I?
Hint #1 _____
Hint #2 _____
Hint #3 _____

What am I?
Hint #1 _____
Hint #2 _____
Hint #3 _____

What am I?
Hint #1 _____
Hint #2 _____
Hint #3 _____

What am I?
Hint #1 _____
Hint #2 _____
Hint #3 _____

What am I?
Hint #1 _____
Hint #2 _____
Hint #3 _____

What am I?
Hint #1 _____
Hint #2 _____
Hint #3 _____

What am I?
Hint #1 _____
Hint #2 _____
Hint #3 _____

What am I?
Hint #1 _____
Hint #2 _____
Hint #3 _____

Ask Me About

Ask Me About

Ask Me About

Ask Me About

Ask Me About

Ask Me About

Ask Me About

Ask Me About

Ask Me About

Ask Me About

Ask Me About

Ask Me About

_____'s Filmstrip

_____'s Filmstrip

_____'s Filmstrip

_____'s Filmstrip

_____'s Filmstrip

_____'s Filmstrip

_____'s Filmstrip

_____'s Filmstrip

_____'s Filmstrip

_____'s Filmstrip

_____'s Filmstrip

_____'s Filmstrip

_____'s Filmstrip

_____'s Filmstrip

What If...?

What If...?

What If...?

What If...?

What If...?

What If...?

What If...?

What If...?

What If...?

What If...?

What If...?

What If...?

What If...?

What If...?

What If...?

Book Review
of _____

Book Review
of _____

Book Review
of _____

Book Review
of _____

Book Review
of _____

Book Review
of _____

Time Line
's _____

Time Line
's _____

Time Line
's _____

Time Line
's _____

Time Line
's _____

Time Line
's _____

Time Line
's _____

Time Line
's _____

Time Line
's _____

Time Line
's _____

Time Line
's _____

Time Line
's _____

Time Line
's _____

Time Line
's _____

Time Line
's _____

Skip Counting
by _____

Skip Counting
by _____

Skip Counting
by _____

Skip Counting
by _____

Skip Counting
by _____

Skip Counting
by _____

Skip Counting
by _____

Skip Counting
by _____

Fact Family

Fact Family

Fact Family

Fact Family

Fact Family

Fact Family

Fact Family

Fact Family

-s Frog

-s Frog

-s Frog

-s Frog

-s Frog

-s Frog

-s Frog

-s Frog

-s Frog

-s Frog

-s Frog

-s Frog

-s Garden

-s Garden

-s Garden

-s Garden

-s Garden

-s Garden

-s Garden

-s Garden

-s Garden

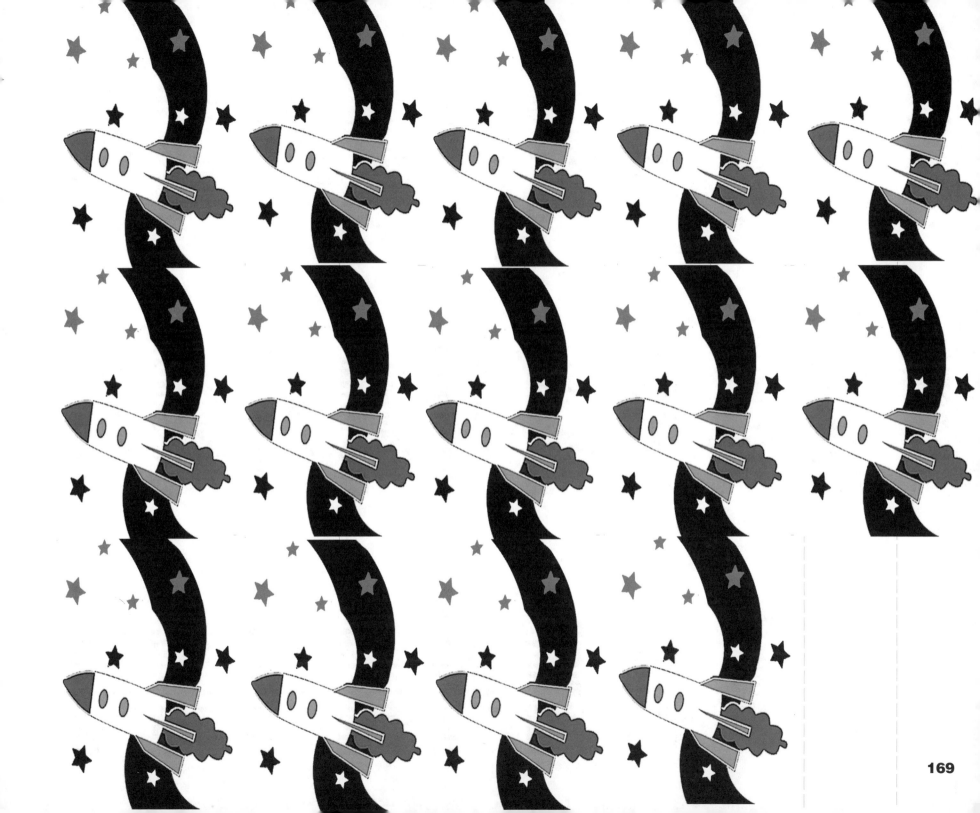

Smartie Putty's

Smartie Putty's

Smartie Putty's

Smartie Putty's

Smartie Putty's

Smartie Putty's

Smartie Putty's

Smartie Putty's

Smartie Putty's

Smartie Putty's

Smartie Putty's

Match the Sound

Match the Sound

Match the Sound

Match the Sound

Match the Sound

Match the Sound

Match the Sound

Match the Sound

Match the Sound

Match the Sound

Match the Sound

Match the Sound

Match the Sound

Match the Sound

Match the Sound

Match the Sound

Match the Sound

Match the Sound

Buddy 's

Buddy 's

Buddy 's

Do Not Disturb!

Do Not Disturb!

Do Not Disturb!

Buddy 's

Buddy 's

Buddy 's

Do Not Disturb!

Do Not Disturb!

Do Not Disturb!

Buddy 's

Buddy 's

Buddy 's

Do Not Disturb!

Do Not Disturb!

Do Not Disturb!

Spring Has Sprung

Spring Has Sprung

Spring Has Sprung

Spring Has Sprung

Spring Has Sprung

Spring Has Sprung

Spring Has Sprung

Spring Has Sprung

Spring Has Sprung

Spring Has Sprung

Spring Has Sprung

Spring Has Sprung

LET ME HOLD YOUR CHANGE

LET ME HOLD YOUR CHANGE

LET ME HOLD YOUR CHANGE

LET ME HOLD YOUR CHANGE

LET ME HOLD YOUR CHANGE

LET ME HOLD YOUR CHANGE

LET ME HOLD YOUR CHANGE

LET ME HOLD YOUR CHANGE

's Play Clay

's Play Clay

's Play Clay

's Play Clay

's Lip Gloss

's Lip Gloss

Finger Paint

Finger Paint

Finger Paint

's Play Clay

's Play Clay

's Play Clay

's Play Clay

's Lip Gloss

's Lip Gloss

Finger Paint

Finger Paint

Finger Paint

's Play Clay

's Play Clay

's Play Clay

's Lip Gloss

's Lip Gloss

Finger Paint

Finger Paint

Finger Paint

Stamp Dispenser

Stamp Dispenser

Stamp Dispenser

Stamp Dispenser

's Pillbox

's Sunscreen

's Sunscreen

Wish Jar

Wish Jar

Portable Aromatherapy

My Traveling Powder

Address Labels

Address Labels

Stamp Dispenser

's Pillbox

's Sunscreen

's Sunscreen

Wish Jar

Wish Jar

Portable Aromatherapy

My Traveling Powder

Address Labels

Address Labels

Stamp Dispenser

's Pillbox

's Sunscreen

's Sunscreen

Wish Jar

Wish Jar

Portable Aromatherapy

My Traveling Powder

Address Labels

Tacks

Tacks

Tacks

Tacks

Rubber Bands

Rubber Bands

Rubber Bands

Rubber Bands

Paper Clips

Paper Clips

Paper Clips

's

_____

Body Glitter

's

_____

Body Glitter

's

_____

Body Glitter

's

_____

Body Glitter

's

_____

Body Glitter

's

_____

Body Glitter

I'll Hold It for You!

I'll Hold It for You!

I'll Hold It for You!

I'll Hold It for You!

I'll Hold It for You!

I'll Hold It for You!

I'll Hold It for You!

I'll Hold It for You!

I'll Hold It for You!

I'll Hold It for You!

I'll Hold It for You!

Hair Kit

Friends Forever

Friends Forever

I've been kissed by a frog!

I've been kissed by a frog!

Toothbrush Holder

Toothbrush Holder

Toothbrush Holder

Fishing Stuff

Fishing Stuff

Hair Kit

Friends Forever

Friends Forever

I've been kissed by a frog!

I've been kissed by a frog!

Toothbrush Holder

Toothbrush Holder

Toothbrush Holder

Fishing Stuff

Fishing Stuff

Hair Kit

Friends Forever

Friends Forever

I've been kissed by a frog!

I've been kissed by a frog!

Toothbrush Holder

Toothbrush Holder

Toothbrush Holder

Fishing Stuff

175

_____'s Cheering for You

_____'s Cheering for You

Thinking Lotion

Dependable

Loyal

Honest

NUTS & BOLTS

Considerate

Responsible

COOL TUBE

First Aid

Home of the "Fit"

EMERGENCY SEWING KIT

SPREADER

GLUE

Animals

Fruits

Sports

Storage

Storage

Storage

Word Storage

Shape Words

Color Words

Guess My Letter

Guess My Letter

Guess My Letter

Guess My Letter

Measure UP!

Piggy in the Mud

Piggy in the Mud

Piggy in the Mud

SORT

SORT

12345 Math Helper

12345 Math Helper

Estimate #3

Numeral – Word Match

Piggy in the Mud

Piggy in the Mud

Piggy in the Mud

SORT

SORT

DICE Probability

DICE Probability

Estimate #2

Letters

Piggy in the Mud

Piggy in the Mud

Piggy in the Mud

SORT

SORT

Build a Pattern

Build a Pattern

Estimate #1

MAKE A BOX GAME

MAKE A BOX GAME

Pennies

Beans

Cross It Out

Cross It Out

Invisible INK

Invisible INK

YELLOW

YELLOW

MAKE A BOX GAME

MAKE A BOX GAME

Popcorn

Cheerios

Penny Toss

FRACTION FUN

Invisible INK

Invisible INK

BLUE

BLUE

Estimate #4

Coin Count

Rice

Cotton Balls

DICE GAME

DICE GAME

DICE GAME

DICE GAME

RED

RED

Which Things Are MAGNETIC ?

Moldy Cheese

Search for the Stars #1

Search for the Stars #2

Search for the Stars #3

Search for the Stars #4

Search for the Stars #5

Rainbow Crayon

Rainbow Crayon

Puffy Paint

Which Things Are MAGNETIC ?

Moldy Cheese

Search for the Stars #6

Search for the Stars #7

Search for the Stars #8

Search for the Stars #9

Search for the Stars #10

Rainbow Crayon

Rainbow Crayon

Puffy Paint

Which Things Are MAGNETIC ?

Moldy Cheese

Search for the Stars #11

Search for the Stars #12

Search for the Stars #13

Search for the Stars #14

Search for the Stars #15

Search for the Stars #16

Puffy Paint

Puffy Paint